ENVIRONMENTAL SCIENCE, ENGINEERING AND TECHNOLOGY

SUSTAINABILITY

INTEGRATING ENVIRONMENT, AGRICULTURE AND RENEWABLE ENERGY FOR FOOD SECURITY

Environmental Science, Engineering and Technology

Additional books in this series can be found on Nova's website
under the Series tab.

Additional e-books in this series can be found on Nova's website
under the e-book tab.

ENVIRONMENTAL SCIENCE, ENGINEERING AND TECHNOLOGY

SUSTAINABILITY

INTEGRATING ENVIRONMENT, AGRICULTURE AND RENEWABLE ENERGY FOR FOOD SECURITY

TOFAEL AHAMED
RYOZO NOGUCHI
AND
TOMOHIRO TAKIGAWA

nova publishers
New York

Copyright © 2015 by Nova Science Publishers, Inc.

Additional color graphics may be available in the e-book version of this book.

Library of Congress Cataloging-in-Publication Data

Sustainability : integrating agriculture, environment, and renewable energy for food security / editors, Tofael Ahamed, Ryozo Noguchi, and Tomohiro Takigawa (Faculty of Life and Environmental Sciences, University of Tsukuba, Japan).
pages cm. -- (Environmental science, engineering and technology)
Includes index.
ISBN 978-1-63117-702-6 (hardcover)
1. Sustainable development. 2. Sustainable agriculture. 3. Renewable energy sources. 4. Food security. I. Ahamed, Tofael, editor.
HC79.E5S86653 2015
338.1'9--dc23
2015000084

Published by Nova Science Publishers, Inc. † New York

CONTENTS

FOREWORD

I was delighted to receive the invitation to write a foreword for this book by Tofael Ahamed, Ryozo Noguchi and Tomohiro Takigawa. It is an ambitious book and, as you will see from a cursory review of the list of contents, a 'book of bits' with major concepts being introduced on nearly every page. The ambition and diversity of numerous elements of this book are linked and central to the book's contribution.

The landscape of sustainability is complex, often confusing and highly contested. These realities are not avoided within the pages of this book. All professionals working in this field have experienced the complexity of sustainability. What is it? How can it be measured? What do the measurements mean and how are they interpreted? Many of us will admit to being confused by sustainability. What is the link between local and global sustainability? How can quantitative and qualitative assessments be combined? How do we manage the hypocrisy we all feel as we compare our unsustainable lifestyles (including attending international conferences dealing with sustainability issues such as global warming) and our well-intended pronouncements on how others should live their lives? In addition, how do we manage the contested views of experts and lay populations regarding the interpretations of sustainability assessments? A reduction in fish quota can mean the end of a viable coastal community. An energy subsidy for photovoltaic cells that are invested in by the rich can increase the heating costs for the poorest members of society. Sustainability is not 'nice', it is not easy and it is not simple. Beyond the much quoted anthropocentric eulogies regarding the heritage of future populations of the world, little agreement exists regarding what sustainability means in theory or effect.

This book contributes to our understanding and awareness of the complexity, confusion and contested views of sustainability.

For example, within these pages, you will find discussions about the sustainability of the environment, agriculture, and renewable energy for food security. In addition, you will find issues resulting from environmental and agricultural sustainability indicators (SI), reflective practice and multiple views of sustainability. The book considers sustainability and ethics in agricultural production and the conflict between food and biofuel as the issue pertains to food security. Agricultural sustainability is demonstrated to face challenges among threats of lacking soil and water recourses, climate change and population growth. Sustainability and system dynamics approaches are discussed, including human development indices. Means to integrate social, economic and ecological indicators to achieve composite sustainable development index are discussed as well as systemic and systematic analysis of sustainability through system dynamics approaches and the changes in our thinking which emerges from making use of backcasting approaches. Further, agricultural informatics are discussed. In this regard ICT is assessed – it seems to encourage sustainable growth and easy access to information throughout the agrarian community with the drive to implement data to knowledge as a part of site-specific management. But how do all these things fit together and how do they make sense? This is where you, the reader are key to the books purpose.

The ambition of the book is clearly demonstrated in the range and scale of issues, tools and ideas covered. This book is a brave attempt to present complexity, to be honest about the diversity of views and present the reader with a compendious overview of the field – and the opportunity to make your own sense of the field. Whether it succeeds or not depends on your tenacity and on your willingness to find and apply what is needed for you in your unique context.

Simon Bell

1st January 2015, Norfolk, UK.

(Simon is Professor of Innovation and Methodology in the Faculty of Maths, Computing and Technology at the Open University in the UK)

PREFACE

This book aims to present sustainability issues in a simple way to students in the field of agriculture and the environment. Viewpoints regarding the meaning of sustainability are diverse. Sustainability must be operational in each specific context and scaled, and appropriate methods must be designed to achieve long-term goals for the environment, agriculture, energy, and food security. The ultimate goal for the environment and agriculture is to protect soil and water to ensure food security for the growing population. This book discusses the integration of views pertaining to the sustainability of agriculture, the environment, renewable energy and food security. Considering sustainability in a broad sense allows us to compare and contrast facts for application across these domains and apply lessons from one to another.

In Chapter 1, a framework is considered that links sustainability and sustainable development. In this section, the most important issues are introduced with the challenges of sustainability analysis to improve agricultural yield, bioenergy resources and major factors that influence the sustainability indices of agriculture. Environmental issues and food security have resulted in interest in biofuel production. A conflict will occur between food, fuel and land use, especially because food security will be a challenge in the coming decades. The ICT and its implications have focused on the most recent developments with the slogan of 'smart technologies all over the world for premium agriculture'.

In Chapter 2, sustainability indicators for ecology, the environment and agriculture are discussed with a vision of reflective practice and from multiple views of sustainability. This chapter reviews some of the core points from the book "Sustainability Indicators, Measuring the Immeasurable?" by Simon Bell and Stephen Morse. In addition, we note the transdisciplinary views of sustainability to develop the project over time. Sustainability must be operational in all contexts, and sustainability indices must play a significant role in measuring sustainability. In this context, Chapter 2 highlights sustainability indicators, such as Environmental Sustainability Indicators, Maximum Sustainable Yield and AMOEBA, Urban sustainability and stakeholders' perceptions of institutional sustainability. Local food-based AMOEBA diagrams include factors that influence the sustainability of the mentioned products. This chapter introduces and discusses systemic alternative approaches used for solving problems for a variety of topics, such as changes in thinking. The demise of narrow scientism is discussed according to different research contributions. The ranges of systemic approaches that help change perceptions and can be considered as paradigm shifts are reviewed. From these systems, logical frameworks are used to represent local food sustainability.

In Chapter 3, challenges in sustainability and the ethics of agricultural production are the main areas of focus, including recent issues in food and biofuel production. In contemporary discussions on sustainability forums, ethical issues in agriculture and food production are confronted and environmental concerns must be considered. Intensive cultivation results in soil quality degradation, fresh water extraction and agricultural chemicals, especially dissolved phosphorus, cumulatively affect the soil. This problem is a worldwide issue. Organic agriculture has become popular within the last few years, with costs and production that are favorable for meeting food security standards. Growing concern regarding the chemical contamination of food has prompted interest in ethics, health and the demand for food security. We are working on food security from one aspect, and working to meet the needs of ecology, agriculture and the environment from another aspect. A weak vision of sustainability focuses on food and energy security rather than sustainability. The major factors in food and biofuel production are briefly reviewed, including three generations of biofuels. The ethical considerations of agricultural operations that lower biodiversity, soil quality and water quality are also included in this chapter.

In Chapter 4, the environmental and agricultural resources utilization are the key area of discussion to ensure the sustainability of food security. In this chapter, we focused on food security and regional indices related to vulnerability and food production in climate change scenarios. The food security and climate change perspectives, which represent the greatest concern to agriculture and the environment, are areas of focus. Sustainable food production and feeding an overburdened population with limited natural resources will be the greatest challenge faced by mankind. This chapter focuses on developing a food system model and food security dimensions with the systematic policy planning that is required to identify the indicators and stakeholders related to food production. Climate change and food security have become a global problem. Regions without food security would suffer human migration towards areas with food security. Food security dimensions, vulnerability and climate modeling tools are briefly reviewed in this chapter. In addition, food insecurity factors and system models are discussed, including basic food security indices related to national resilience, household access, GDP, and food nutrition.

In Chapter 5, a systematic analysis of good agricultural practices and indicators is presented. A systematic analysis using indicators for sustainability and system dynamics approaches is integrated to show life and human development affluence indices and ensure quality of life in sustainable development. To consider the time scale, system dynamic models are useful for understanding the behavior of complex systems over time. In this chapter, the indicators for the ecological, economical and human development indices are discussed. A system dynamic model that connects all of the indicators is shown to demonstrate their positive and negative impacts on the human development index.

In Chapter 6, sustainability and changes in thinking are discussed using backcasting approaches. This chapter introduces and discusses systemic approaches for thinking and problem solving over a wide range of topics, specifically ecology and energy modeling. A variety of methods can be used to approach systemic problem solving, including the introduction of backcasting to understand future needs. This chapter uses examples and tools that are commonly practiced by researchers and policy planners. The system approach is required to ensure the sustainability of the projects, institutions and programs over time. System tools and theory are discussed, as well as neural networks and genetic algorithms.

Chapter 7 discusses energy systems, including energy use, to consider food production based on system informatics. Life Cycle Assessment (LCA) refers to a "cradle to grave" analysis that is used to estimate the total costs of developing a product from the beginning to the end of its lifecycle. The system design tool discussed in this chapter uses linear programming as a standard method for obtaining an optimum solution to maximize or minimize the objective function. The most important part of the energy system may be renewable energy. In addition, biofuel production systems are evaluated under different scenarios, and the decision support systems can be utilized for numerical simulation, modeling and data mining. This chapter also presents a spatial decision support system for determining renewable energy potentials.

Chapter 8 focuses on sustainability and informatics, which broadly discuss ICT issues that could serve as a gateway for accessing and sharing information for agricultural sustainability. The innovation of smart applications and precision agriculture tools ensures the site-specific management of crops over the growing season. Furthermore, innovation in mobile technologies and Big Data issues are also discussed. Big Data and mobile technologies that refer to smart applications are introduced in this section. Informatics considers this multidisciplinary field in terms of technology transformation and its influences on our life. Our focus in this chapter is on visualizing the roles of informatics and smart technologies in agricultural sustainability and food security issues in the coming decades.

Overall, this book focuses on discussing aspects of sustainability in simple and pragmatic ways for readers. Different debatable issues and factors that influence the site-specific management of agriculture and the environment are discussed. Sustainability must be operational in each specific context at scales that are appropriate for its achievement, and appropriate methods must be designed for its long-term measurement. This book integrates views of sustainability for agriculture, the environment and renewable energy. Understanding sustainability in a broad sense allows us to compare and contrast facts of application across these domains and to apply lessons from one to another. Many individuals have noted the need for measuring sustainability. This book discusses a few approaches for achieving this goal and their problems. The key issues highlight sustainability as a reflective practice, consider multiple views of sustainability, sustainability in agriculture, ethics in agricultural production and the conflicts between food and biofuels in terms of food security. Furthermore, this book focuses on the systematic analysis of sustainability using system dynamics approaches. Sustainability and informatics that provide a gateway for accessing and sharing agricultural sustainability information are required to connect the parameters that lead to sustainable development. Informatics considers this entire multidisciplinary field for the transformation of technology. This book has attempted to integrate these perceptions.

Tofael Ahamed
Ryozo Noguchi
Tomohiro Takigawa

ACKNOWLEDGMENTS

This book is not only a product of the efforts of the authors but also has benefited from many researchers who have helped complete it. We were inspired by the course "Concepts of Sustainability Indices" offered in the Master's Program in the Agro-bioresources Sciences and Technology at the University of Tsukuba. Over the last four years, we have taught from the books of Simon Bell and Stephen Morse, which inspired us to extend the sustainability concept further by integrating the environment, agriculture, and renewable energy with perspectives of science and engineering for food security.

We are grateful to Professor Simon Bell from Open University, UK for contributing the foreword of this book, Professor Prasanta K. Kalita from University of Illinois at Urbana-Champaign, USA, and Hiroyuki Konuma, Regional Representative for Asia and the Pacific, Food and Agriculture Organizations of the United Nations for comments about the contents. We also extend our sincere appreciation to Professor KC Ting, University of Illinois at Urbana-Champaign and to Professor Erdal Ozkan, Ohio State University, USA for their compliments.

Our sincere appreciation and gratitude extend to the University of Tsukuba, Japan for providing assistance and support for publishing the book. We thank Nazia Muhsin, Riska Ayu Purnamasari, Afsana Akther, Summaya Salman, and Shah Alamgir for their help in completing the manuscript. We offer special gratitude to the graduate students Nazia and Riska for drawing many of the figures in this book and for checking the references thoroughly. We could not have finished this book without their continuous support and help.

Furthermore, we extend our thanks to Mr. Yano Shuhei for drawing the system dynamic models for determining the maximum sustainable yield and development indices. We express our thanks to students in the Master`s Program at the Graduate School of Life and Environmental Sciences at the University of Tsukuba, Japan, for their participation in discussions regarding the content of the manuscript and its clarity for students.

We extend appreciation and special thanks to the support staff, Ms. Yoshimaru Shiori of the University of Tsukuba. Last but not least, we extend sincere appreciation and gratitude to our family members for their continuous support of our efforts to complete this book for the benefit of students and researchers.

Chapter 1

SUSTAINABILITY AND SUSTAINABLE DEVELOPMENT: A SUSTAINABILITY PERSPECTIVE ON THE INTEGRATION OF ENVIRONMENT, AGRICULTURE, AND BIOENERGY FOR SUSTAINABLE DEVELOPMENT

A debate regarding the cross-cutting points between sustainability and sustainable development has occurred within the last decades. These two interdisciplinary issues and their explanations can be discussed separately. Sustainability refers to maximizing the net benefits of economic development to maintain services and the quality of natural resources over time (Pearce and Turner, 1990). Sustainable development focuses on development that meets the needs of the current generation without compromising the ability of future generations to meet their needs and aspirations (WCED, 1987). The focus of this chapter is to review a few models that describe sustainability within five categories of models: quantitative (mathematical, statistical, database, econometric, and computer simulation), pictorial or snapshot (Venn diagram and graphical representation), conceptual models (concepts and theories), standardizing (indicators and benchmarks) and physical (Todorov and Marinova, 2010). Our aim is to highlight sustainability and system approaches throughout all chapters. Using this connection, the dynamic modeling of sustainable development has been reviewed based on sustainability indicators. First, the five most important methodological criteria that must be considered to model the issues of sustainability must be discussed, an interdisciplinary approach, uncertainty management, a long run or intergenerational viewpoint, the global-local perspective and the stakeholder perception (Boulanger and Brechet, 2005). Recently, the transdisciplinary concept for the sustainability of a project has been considered widely. We have attempted to identify and connect the indicators for measuring sustainability and sustainable development.

1.1. MEASURING SUSTAINABILITY AND SUSTAINABLE DEVELOPMENT INDEX

Measuring sustainability and sustainable development is a challenge that depends on social, economic, and ecological indicators. Almost all indicators in the environment are connected. These indicators constitute a matrix for gauging sustainable development and can be expressed as the Global Competitiveness Index, Environmental Sustainability Index and

Human Development Index (Zgurovsky, 2007). If we focus on food, water, population, urbanization, energy, poverty, health and climate change, we can link each of these indexes and correlate them with one another using the indicators of economic, ecological, and social aspects in different dimensions (Figure 1.1).

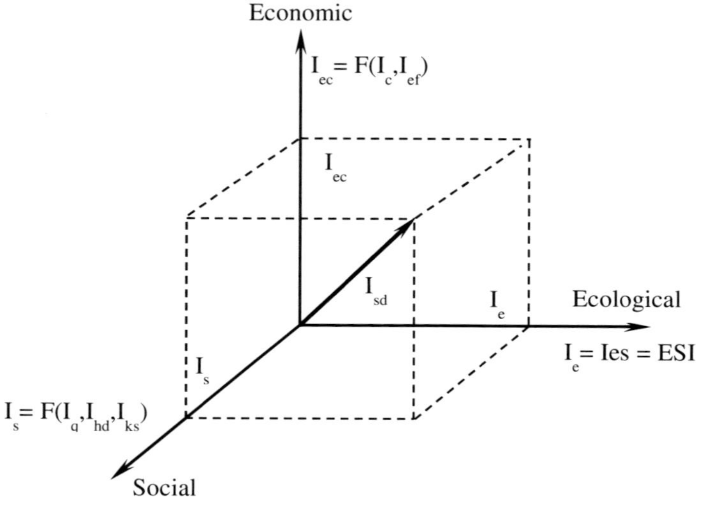

Figure 1.1. The dimensions for gauging the sustainable development matrix considering economic, social, and ecological aspects (Zgurovsky, 2007).

The sustainabilities of these dimensions are methods that can be used to link these indicators with other indicators, resulting in sustainable development. The sustainable development model is represented mathematically as the sum of the security of a population and its quality of life, where quality of life is a sustainable development component that consists of economic, ecological, and social dimensions (Zgurovsky, 2007). The generalized sustainability measure is presented using a quaternary $\{Q\}$, with an imaginary scalar part, j, that describes security and quality of life in three dimensional spaces:

$$\{Q\} = jI_{sec} + \overline{I_{ql}} \tag{1-1}$$

where $j = \left\{\frac{1}{\sqrt{-1}}\right\}$ for real positive values of I_{sec} greater than 0. In the spatial scale, the sustainable radius vector (I_{sd}) with the Euclidean norm is a complex number whose modulus or absolute value has been reported by Zgurovsky.

$$I_{sd} = \|\overline{I_{sd}}\|\sqrt{(I_{ec}^2 + I_e^2 + I_s^2)} \tag{1-2}$$

Here, I_{sd} is the sustainable development index and I_{ec}, I_e, and I_s are the economical, ecological, and social indicators, respectively. Again, I_{ql} is the quality of life component that results from the projection vector of I_{sd} on the ideal vector with coordinates (1:1:1). The quality of life component can be expressed as follows:

$$I_{ql} = \sqrt{I_{ec}^2 + I_e^2 + I_s^2}.cos\partial \quad (1\text{-}3)$$

Here, ∂ is the degree of harmonization, which can be expressed as follows:

$$\partial = arccos\left[\frac{I_{ec}^2+I_e^2+I_s^2}{\sqrt{3}\sqrt{I_{ec}^2+I_e^2+I_s^2}}\right], 0 \leq \partial \leq arccos\left[\frac{1}{\sqrt{3}}\right] \quad (1\text{-}4)$$

The quantitative methodology of sustainable development from Zgurovsky's method is a benchmark for system dynamic models. The main task for measuring sustainable development is to develop a scale of indicator values and test the quality of the indicators. One of the most important indicators is the 'Gross Environmental Sustainable Development Index,' which attempts to describe the quality indicators.

1.2 Gross Environmental Sustainable Development Index (GESDI)

The GESDI emphasizes the collection of all effective aspects related to the daily life of humans and communities that significantly influence social, economic, health, physical, biological, and cultural elements (Global Community, 2004). From this perspective, complete data regarding the status and deviations of the variables are required to present an effective measurement of quality. However, these data are expected to cluster based on their relations with indices, which are used to speculate the effects of public and private actions. These indices also measure the efficiencies of various actions from all aspects. Again, importance is given to achieving reliable data for measuring the impacts of human activities on the environment. In addition, it is equally important to determine the possible activities that would help to mitigate the negative impacts of human activities. The well-being of a community also reflects the quality of the urban environment. In contrast, urban environments depend on adequate food supply, the availability of proper nutrition, health care systems, and educational systems, the development of science, research and technology, senior citizen care, food supply systems, population growth rate, infrastructure, recreation facilities, roads and transport infrastructure, pollution, and waste disposal policies. All of these components of human life help to identify indicators of different aspects, such as economic data for an individual person, family and household, the status of the country's economy, the region's economy, social quality, and housing and community facilities. Weights can be placed on different elements of the measurement by using assumptions from the survey results using a values scale. An impact or stress equation can be defined to determine the GESDI (Global Community, 2004). The equation itself divides the ecological balance and pollution into core sustainable development impacts. The four basic quality systems (denoted as *I*) include people, economic development, environment and the availability of resources and their interactions that form an impact or stress (Figure 1.2).

The function *I* is a product of the four index scales of assessments, *U*, *G*, *P*, and *C*. Each product indicates the relative importance of a given impact or stress with respect to the four major quality systems.

$$I_i = I_{i,E} + I_{i,AR} + I_{i,PA} + I_{i,ED} \quad (1\text{-}5)$$
$$I_{i,E} = U_i G_i P_i (WC)_{i,E} \quad (1\text{-}6)$$
$$I_{i,PA} = U_i G_i P_i (WC)_{i,PA} \quad (1\text{-}7)$$
$$I_{i,AR} = U_i G_i P_i (WC)_{i,AR} \quad (1\text{-}8)$$
$$I_{i,ED} = U_i G_i P_i (WC)_{i,ED} \quad (1\text{-}9)$$

Here, *E*, *PA*, *AR* and *ED* represent the components of the four major quality systems, including the economic development processes of individual households or communities and any task in which human performance is related to the components of Economic Development (*ED*). New processes or systems eventually fill this system. However, the main purpose of this method is to find a safe process. Human activities and their actions, interactions and reactions are represented by elements of individuals and social systems (*PA*). The social and economic wellbeing of individuals and their health are the main concerns in this system. However, the impacts are transferred through the medium provided by the components of the environmental system (*E*). Here, each component interacts with the other components and with the components of other systems. According to the subjective point of view, the impact *I* represents the development of that aspect. For example, if only the impact or stress associated with the environment is considered, then I_i is called the 'Environmental Quality Equation,' and the result is 'Sustainable Environmental Development'.

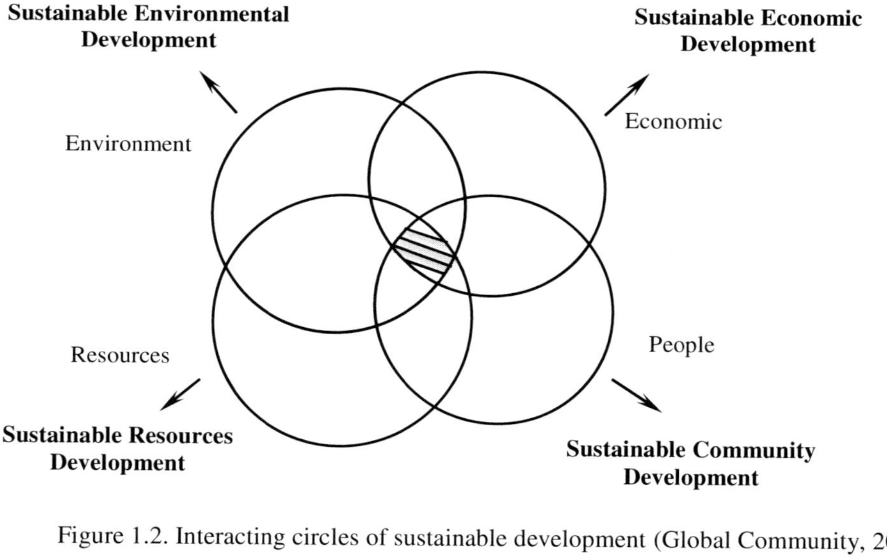

Figure 1.2. Interacting circles of sustainable development (Global Community, 2004).

Similarly, if only the impacts or stresses associated with 'Economic Development' are considered, then the result is called 'Sustainable Economic Development'. In addition, if only the impacts or stresses related with people are considered, then the obtained results are called 'Sustainable Community Development' (Global Community, 2004). The core feature is the interaction of all impacts or stresses with the four major quality systems. In this case, *U* refers to the urgency index, which states the importance of the requirements for finding a solution to a given stress within a reasonable period of time to prevent the impacts from damaging the components of the four major quality systems. Furthermore, *G* refers to the geographical extent index, which is used to identify the significant and detectable geographical extents of the stress or impact. Thus, the equation contains all of the important modes and methods of transportation and communication. In addition, *P* refers to the persistence index, which is used to identify the duration of time during which the effects of stress or impact will be felt at a significant level. The number of interactions with the index is represented by *C* and elaborates the relative complexities of the stress or impact and simultaneously interacts with each of the four major quality systems. The weight (*W*) indicates the possibility of an interaction or its decision value and its level of importance. A set of criteria pertaining to ecological, economic, and social aspects of agricultural sustainability has also been identified (Reijntjes et al., 1992). The ecological criteria comprise the use of nutrients and organic materials, water, energy, and environmental effects. In addition, the economic criteria include farmers' livelihood systems, competition, factor productivity and comparative values of the external inputs for crop yield. Therefore, sustainability in agriculture is a major concern of sustainable development, which includes the above mentioned factors. The following sections briefly describe the focal points in agriculture and natural resources and trade-off factors that must be considered in the coming years for food security.

1.3. SUSTAINABILITY IN AGRICULTURE

Sustainability in agriculture is multidimensional; nevertheless, a number of indicators and parameters have been suggested for measuring agricultural sustainability over time. Agricultural sustainability must be considered on spatial and time scales. Numerous indicators have been developed, but do not cover all dimensions and levels. An ecological footprint is a sustainability case that indicates the amount of productive land used by individuals, cities and countries (Figure 1.3). Therefore, indicators used for agricultural sustainability should be location specific (Hayati et al., 2010). For any sustainable agriculture study, the question of how agricultural sustainability can be measured arises. Some individuals have indicated that the concept of sustainability is a "social construct" (David 1989; Webster 1999).

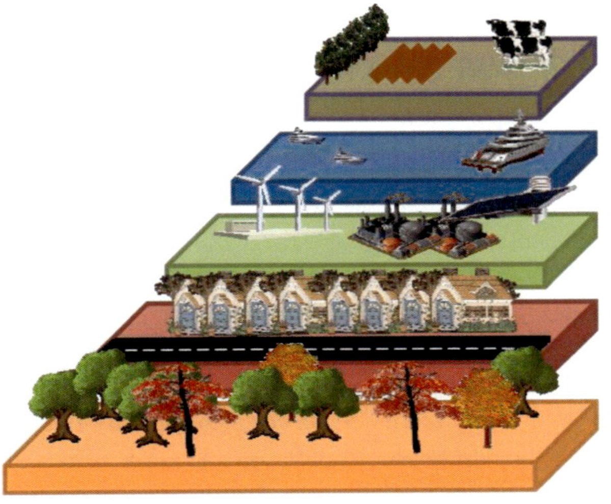

Figure 1.3. Land categories used in the ecological footprint analysis.

It is impossible to precisely measure sustainability because sustainability is a site-specific and dynamic concept (Ikerd, 1993). Recently, the Organization for Economic Co-operation and Development (OECD) developed a common framework called "driving force state response" (DSR) to help to develop indicators. Driving force indicators refer to factors that cause changes in farm management practices and input use. State indicators show the effects of agriculture or the rate of change in the environment, such as soil, water, air, biodiversity, habitat, and landscape. Response indicators refer to the actions that are taken in response to the changing state of the environment. Using the DSR framework, OECD (1997) identified 39 indicators of issues, including farm financial resources, farm management, nutrient use, pesticide use, water use, soil quality, water quality, land conservation, greenhouse gases, biodiversity, landscape, and wildlife habitats, and presented different contextual farming information, including socioeconomic background, land-use and output. Most of these indicators could be used to identify agricultural sustainability at the aggregate level. The evaluation process for the sustainability of rice production in India used nine indicators, including integrated nutrient management, land productivity, integrated water management, integrated pest management, input self-sufficiency, crop yield security, input productivity, information self-reliance and family food sufficiency (Gowda and Jayaramaiah, 1998). In the productive agricultural policy, long-term crop yield trends could potentially increase the biological productivity capacity of agricultural lands and the ability of agriculture to sustain its resource production capacity.

However, because of food security issues and high yield expectations, fertilizers are used to increase yields. In some cases, the overuse of fertilizers on fields cause soil degradation and contamination with agricultural chemicals. Additionally, excessive fertilizer use can cause eutrophication, acidification and the toxic contamination of soil, water and air. Agriculture nutrient balance parameters include the gross nutrient balance and input and output ratio. The gross nutrient balances of the total nitrogen, phosphorus and potassium (NPK) concentrations result from the application of chemical fertilizers and livestock manure. Consequently, these nutrients become inputs through irrigation, rain and biological fixation, and the total nutrient input can be calculating by subtracting the nutrients lost through absorption by agricultural plants, run-off, leaching and volatilization from these inputs. Soil quality indicators are additional factors that include physical properties, such as the soil texture, depth, bulk density, water holding capacity, water retention characteristics, and water content. The chemical properties include the total organic C and N contents, organic matter content, pH, electrical conductivity, mineral N, extracted P, and available K, and the biological properties include the microbial biomass C and N, potential mineral N, soil respiration, total organic C ratio, respiration and the biomass ratio. The type of fertilizer and irrigation system affects the efficiency of fertilizer, pesticide and water use. Agro-environmental indicators provide information regarding the environmental impacts of the production process. The degrees of soil degradation and water pollution are listed. The effects of water and wind erosion, salinization, acidification, toxic contamination, compaction, water logging, and decreasing levels of soil organic matter determine the degree of soil degradation. The quality of surface, ground and marine water is calculated from the nitrogen, phosphorus, dissolved oxygen, toxic pesticide residue, ammonium and suspended sediment concentrations (in weight per liter of water). Soil degradation negatively affects the biodiversity of plants and livestock in agricultural production, which is important for conserving the agro-ecosystem balance. However, the dependence of agricultural production on a limited number of varieties and breeds may increase the susceptibility of crops to pests and disease. Biodiversity measurements reflect the total numbers of varieties/breeds used for major crop/livestock production and the numbers of animals and microorganisms that exist during production. Aspects of sustainable agriculture and the sustainability of agro-ecosystems are reflected in environmental factors and in economic soundness and social considerations. These aspects are observed as net outputs and are observed as changes in the levels of managerial skills among farmers and the income and farming practices of land managers. Few issues are now considered as a primary concern for achieving agricultural sustainability and food security. It is important to understand the parameters that are related to one another and, in most cases, act as constraints on agricultural sustainability and food security. Several factors that act as constraints for achieving sustainability are noted.

1.4. Population and Agricultural Sustainability

Population growth causes pressure on resources, the environment and social systems. Humans are the essence of economic and social systems and the motivating force and ultimate beneficiaries of development. It is generally believed that the world population significantly expanded during the 18th century. In the second half of the 20th century, population growth reached unprecedented levels during the second technological revolution, which resulted in the development of social productivity. From the perspective of resources, increasing populations inevitably result in increasing demands, which leads to excessive resource consumption and enhances the resource crisis. From an environmental perspective, population is increasing rapidly and has resulted in increasing resource utilization and ecosystem damage. Due to increased production and consumption, population groups will encounter land desertification, forest and grassland destruction and the resulting greenhouse effect. If population growth occurs rapidly, the agricultural labor surplus problem will be exacerbated in developing countries, which will accelerate the aggregation and migration of populations and result in decreasing urbanization quality, increasing unemployment and a series of social problems, such as public security needs amidst chaos.

The French demographer Southwick introduced a relevant demographic theory regarding moderate populations that stated a moderate number can achieve the maximum efficiency of the social population. Thus, the expansion of population reproduction achieves a minimum, fertility is restricted, the population structure is coordinated and people have longer life expectancies, which constantly improves the quality of the population. For sustainable development, it is important to improve the quality of the population while simultaneously controlling the population number. With the development of society and advancements in agricultural modernization, the agricultural population will continue to decrease. However, the requirements of modern agriculture and the agricultural labor force will continue to increase. Therefore, increasing human capital accumulation is extremely important to the sustainable development of agriculture.

The accumulation of human capital is useful for improving income distribution. The main cause of poverty among farmers is not the amount of land they possess but whether their accumulation of human capital is too small. Improving the accumulation of human capital will improve farmer income. The most fundamental measure is a change in the income distribution conditions. The first cause of this change is that human capital yields are much greater than physical capital yields, which reduces the share of national income asset contributions and increases the share of labor contributions. Second, because there are many ways to accumulate human capital, the allocation of the crowd is more important than the allocation of physical capital or the allocation of existing assets. Third, human capital promotes the escape of the agricultural labor force from agriculture reductions due to the pressure of the agricultural population on natural resources.

1.5. Natural Resources and Agricultural Sustainability

Natural resources refer to things that enhance human survival, development and enjoyment, such as land resources, water resources and mineral resources. For agricultural production, natural resources are critically important. Natural resources are a basic factor in the development of the agricultural economy. With increasing population and living standards, society requires more agricultural products, which forces people to improve the use of natural resources and impacts natural resources and conflicts pertaining to the sustainable development of agriculture. Sustainable agricultural development is necessary for continued survival and development. The use of agricultural natural resources must consider current interests and the needs of future generations for development. Natural resources support sustainable human development. Their bearing capacity indicates that the size of the world and the number of people in the world restrict the size of the human economic and social structure. As agricultural processes are modernized, continuous pressure is placed on external resources. On the temporal and spatial scales, the overall quality and quantity of natural resources in agriculture are narrowing.

1.6. Climatic Change and Agricultural Sustainability

Climate change is widely recognized as one of the most complex challenges that will be faced in the coming decades. Agriculture is likely to suffer from the impacts of climate change, and sound adaptation of agricultural technologies is required to sustain agricultural production and food systems. Crops are sensitive to temperature variabilities. Each crop has a range of suitable growth temperatures. In addition, because temperatures that are too high or low are adverse to crop growth, climate change may result in a total production loss. Climate change increases the probability of drought, flooding and other extreme weather events. To some extent, climate change may lead to further water resource shortages. Climate warming could result in the expansion of agricultural pests, which would affect crop growth. From the perspective of agricultural production, winter temperatures have an important influence on insect outbreaks. Specifically, cold winters can freeze pests effectively and warmer winters can provide opportunities for pest outbreaks. Climate change results in the deterioration of the agricultural ecological environment. In addition, climate change has played an important role in keeping smallholders at the bottom of value chains. Farmers can no longer rely on timeworn coping strategies when all of their familiar benchmarks for making agricultural decisions (such as the timing of rains for planting and pasture, the probability of frost, the duration of dry intervals that spare crops from disease) are decreasingly reliable. Severe and unexpected weather is shrinking already-limited yields and promoting migration from rural areas and rural jobs. Weather-related events leave developing-country governments, who lack the resources and private sector investment to provide risk management instruments, to cope with major crop failures and their displaced victims after the events (McNamara et al., 2011). Presently, we can recognize three major categories of information systems that have been developed to study the issue of climate at local to regional levels: comprehensive systems and methodologies for institutions, downscaling tools for working at the national and sub-national levels and systems and tools for the crop production, forestry, fisheries, and livestock sectors (Sala, 2007). The first category essentially comprises theoretical methodologies based on different assumptions and approaches that are designed to identify and quantify the impacts of climate change and assess vulnerability to climate change (IPCC, 2001). The second category includes all tools needed to produce climatic data at an appropriate scale for impact modeling and scenario development at the local to regional level (e.g., the 'Statistical Downscaling Model', SDSM; the 'Country Specific Model for Intertemporal Climate', COSMIC; the 'Providing Regional Climates for Impacts Studies' tool, PRECIS). Downscaling tools are applied to develop climate information at high resolution by processing global climate models built with General Circulation Models (GCM). These global models cover areas of 150-300 kilometers. The third and final category consists of all information tools through which it is possible to investigate climate change issues within specific sectors, such as the economic, human health, coastal protection, agriculture, water management, and forestry sectors (Sala, 2007).

1.7. Sustainability and Renewable Energy

Environmental sustainability assessments for renewable energy (e.g., biofuels) are difficult to obtain because of the complexity and the multiplicity of indicators. Some of these indicators have a global effect, such as GHG, while other indicators have local or regional effects, such as water management, soil and resource depletion and local pollution (FAO, 2013a). Agriculture not only uses energy, such as sunlight and fossil fuels, but is also part of the energy supplied from biomass production. The energy potential must be analyzed based on geographic and climatic conditions, including real-time weather data, to promote renewable energy utilization using smart grid approaches. Notably, renewable energy could provide the most beneficial solutions for energy security, new employment and sustainable development in rural areas throughout the world. However, large land areas are needed to produce bio-energy crops or other forms of renewable energy in sufficient quantities to satisfy our energy demands. Intense competition may occur between bio-energy crops and food production regarding environmental affairs. It is important to develop methods for estimating the renewable energy potentials of solar, wind, small hydropower and biomass power generation alternatives for energy production.

During recent years, increasing interest in alternative energy and energy conservation has developed because of fuel prices and CO_2 emissions. Policy makers and researchers have made important contributions towards the development of renewable and sustainable energy sources. Renewable energy is increasingly important. Energy sources such as solar, wind, hydropower and biomass have become options for fulfilling energy demands in rural sectors. Simultaneously, technologies for conserving energy and converting feedstocks to biodiesel and ethanol have significantly improved. The renewable energy potential is one of the most essential pieces of background data for the introduction of renewable energy. This assessment is becoming more significant because other renewable energy sources, such as solar, wind, hydropower and biomass, could fulfill rural energy demands. The most important technologies for enabling renewable energy include energy distribution systems rather than nuclear or thermal power, which require energy concentration systems. The introduction of renewable energy projects requires detailed and specific analyses in specific areas. After the renewable energy potential is known, the best alternative must be selected to introduce renewable energy in rural areas.

1.8. Sustainability and Food Security

Food security is a major concern in the context of agricultural sustainability and the sustainable supply of food for the increasing population (Figure 1.4). Approximately, 870 million people are malnourished throughout the world due to lacking amounts of nutritious foods (FAO, 2012). Furthermore, the undernourishment rate throughout the world is increasing (Figure 1.5). One cause of this increase is the use of improper food systems in the supply chain. The food system includes food production, processing, distribution, marketing, acquisition, and consumption. Therefore, sustainability in food systems plays a significant role in ensuring the delivery of nutritious and safe foods that achieve food security. Priorities are necessary for focusing on global environmental changes to meet the challenges of food security in the coming years. The challenges of food security may need to consider soil quality, land use policies, irrigation management, water management, food distribution and allocation, methods for increasing production, reducing post-harvest losses and reducing food waste, and the control of food prices, food habits, population growth, agricultural biodiversity, and adaptation to climate change.

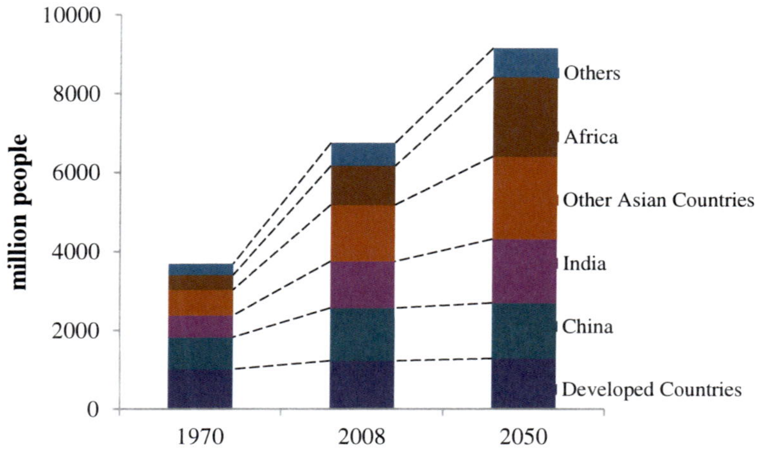

Figure 1.4 Population growth over time and the projected population in 2050 in different regions of the world (UN, 2008).

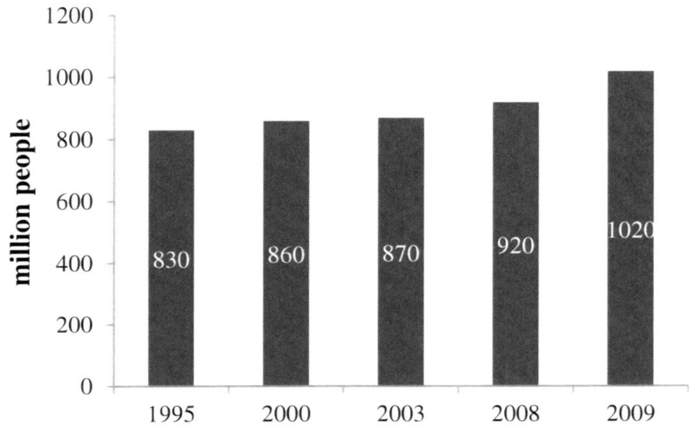

Figure 1.5 The world population suffering from malnourishment or undernourishment (FAO, 2012).

In addition, a healthy, proper, nourishing system includes environmental quality, economic vitality, human health, and social equity. To provide social equity at all levels, a perception or change in thinking is required in society.

1.9. Sustainability and Changes of Thinking

Decision makers must decide policy directions considering agricultural production, processing, transportation and other logistics to ensure food security. Furthermore, energy security is a priority issue for sustainability perspective. While renewable energy is a highly attractive concept, it presents a number of relative weaknesses compared with fossil fuel energy in terms of energy density and stability. Thus, efforts to introduce renewable energy must accurately assess the potential of renewable energy. While engineering approaches offer the opportunity to address this problem as faced in the agricultural field, it is important to discuss this problem and engineering approaches for solving this problem with a variety of individuals who are not engineers. Thus, developing an interface between non-engineers and engineers is an important step in this discussion that can be worked out using holistic approaches.

The holistic approach analyzes the feedback relationship between the parts of a system and the whole system while harmonizing relationships with the stakeholders. In particular, bioenergy is closely related to agriculture, forestry, the environment, and society. Thus, this analysis must consider bioenergy, human thinking and human relationships to maintain sustainable bioenergy production while considering food production and environmental protection. This bioenergy analysis can be used to measure a variety of results, including optimism towards introducing bioenergy policies or pessimism and a recommendation to refrain from introducing a bioenergy policy. Biofuels have attracted attention as high value biomass alternative energy sources that could replace fossil fuels and reduce CO_2 emissions. In conventional systems designed for biofuel production, the overall benefits are considered after calculating the field area and the type of biomass, the characteristics of the biofuel plant and the characteristics of the biofuel based on a single energy use and a single benefit. From the viewpoints of quality of life, environmental protection and sustainable development, biofuel production based on backcasting approaches can be introduced to identify policies and programs that connect a defined desirable future to the present (Oliver and Brooks, 2005). If the absolute benefits are considered in the design of the biofuel production system, then the biofuel types, biofuel plants, biomass types and field areas are ranked based on the application to multiple types of energy and their final benefits.

1.10. SUSTAINABILITY, INFORMATION AND COMMUNICATIONS TECHNOLOGY (ICT)

This discussion obviously established the motivation behind the goal of sustainable agriculture and bio-production by considering the potential environmental and societal benefits. Environmental assessments should focus on soil fertility and structure, water management, and the efficient use of inputs. In addition, ICT can be employed to monitor and manage field conditions, including precision agriculture and intelligent agriculture technologies and energy systems. Precision agriculture relates to monitoring and site-specific management with a geo-referenced decision support system to provide farmers and researchers with technical knowledge to support decisions for optimizing profits (Ahamed et al., 2014). These data are also useful as a reference for researchers who are interested in advancing decision support systems. Intelligent agricultural systems using innovative technology and design can encourage production, increase working efficiency and reduce the amount of heavy labor required of farmers. Energy systems, decision support systems and rural energy simulators can help people monitor the environment and make decisions for growing food or biomass crops (Figure 1.6).

Figure 1.6. ICT for major areas of agriculture, including precision agriculture, automation, and post harvest.

In all phases of the agricultural industry, information technologies are essential for the management and success of businesses. In addition, agriculture has been considerably influenced by ICT in various ways. Many innovations have been made in mobile and energy spaces that could have large impacts on farms. For example, smart power systems, precision agriculture tools, farm management software, data mining techniques, cloud computing, Big Data analysis and processing, and affordable sensors are all currently within the reach of even the smallest farmers. Entrepreneurs are making use of the new technologies that make these products possible. We see potential in radio frequency technologies (RFID, NFC, Bluetooth) and the Big Data that result from them, as well clean energy technology advances that include ambient energy, waste-to-energy, and renewable sources. Big Data have resulted in revolutionary changes in the field of agriculture. Big Data refer to the collection of large and complex data sets that exceed the abilities of commonly used software tools for capturing, curating, managing and processing data within a tolerable period. Experts at the Wisconsin Institute of Discovery (WID) characterized Big Data by using three components, volume (how big the data is), velocity (how fast the data are gathered) and variety (how diverse the collected data are) (Sonka, 2014). Eventually, this Data to Knowledge effort became considered as the Big Data initiative (Figure 1.7). This Big Data concept is an ICT-based innovation and has driven the economic transformation of precision agriculture from a business perspective. The use of scientific data principles, or Big Data technology, in the field of agriculture is productive. The apprehension of the entire farming community regarding the adoption of unfamiliar technologies is well understood. The greatest concern for farmers at this point is to expose their expert knowledge to external analysis.

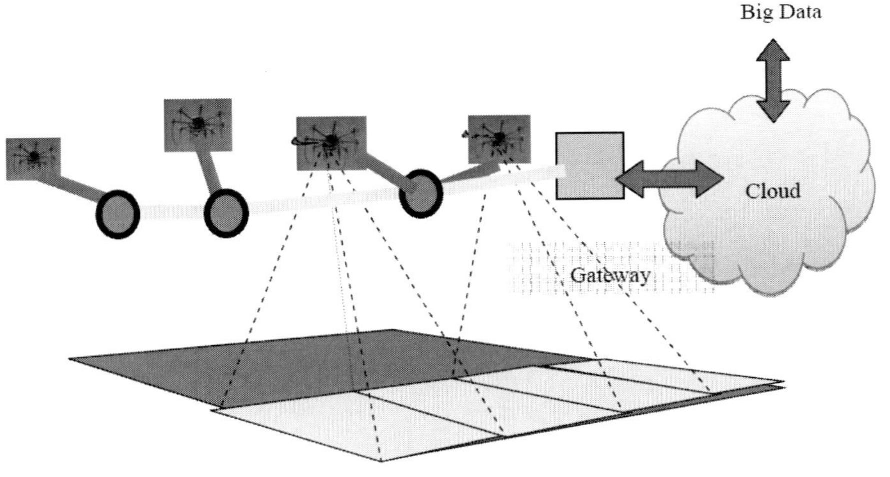

Figure 1.7. Sensor networking and cloud computing for Big Data initiatives.

It is necessary to analyze data to compile important decision-making information for agricultural-related activities. To process large amounts of information, specific techniques are required, such as data mining (sometimes referred as knowledge discovery), which assists the analysis of data from different perspectives and summarizes it into information that is useful for agriculture. Cloud computing represents a new model of network applications was proposed by Google. Today's agricultural industry is embedded with advanced services such as GPS and monitoring sensors that allow for communication throughout the agricultural production system for analyzing and exchanging data. In addition, ICT offers services for agriculture in the form of the cloud. The agricultural cloud offers service expertise to farmers regarding the cultivation of crops, pricing, fertilizers and detailed methods for curing diseases (Miao et al., 2013). Experience with ICT in rural areas suggests that recent technological advances require distinctive approaches for agricultural development. The ability to gather agricultural information through advanced sensing systems has steadily improved within the last two decades, and the unit cost per bit of information has decreased in comparison with the cost decreases in other industries (Ting et al., 2011). Previously, innovations in agricultural machinery have aimed to mechanize agricultural processes. Although this target remains, technological advances currently allow for a broader target. As sensors have become more advanced, ICT can connect with bio-production systems in a variety of ways. The four major areas of ICT involvement are: "precision agriculture", "automation and advancement of intelligent machines," "postharvest process" and "energy systems". These four systems are closely connected and strongly influence sustainable agricultural practices.

Sustainable agriculture seeks and promotes decentralization and the establishment of a networked society. Utilization of renewable energy, attention to food and life quality and the participation of non-farmer citizens in agriculture are considered in sustainable agricultural development. Information technology is known to result in new changes in sustainable agriculture. However, it is equally important for farmers to learn and understand the environmental implications and uses of technologies. The following chapters address further sustainability indicators and the development of indices based on these indicators. Sustainability approaches and system views are integrated and related with environmental, ecological, food and bioenergy issues, including ICT.

Chapter 2

SUSTAINABILITY INDICATORS: REFLECTIVE PRACTICE AND MULTI-VIEWS OF SUSTAINABILITY FOR ECOLOGICAL AND ENVIRONMENTAL MODELING

Sustainability must be operational in all contexts, and sustainability indices should play significant roles in measuring sustainability. In this context, Chapter 2 highlights sustainability indicators, problem-solving approaches using analytical hierarchy programs, ranking systems, and conceptual diagrams. This chapter includes an intensive literature review, especially referring to the book "Sustainability Indicators, Measuring the Immeasurable?" (Bell and Morse, 2008).

2.1. SUSTAINABILITY

Sustainability may be perceived as a complex and difficult concept to understand. Many sustainability experts and researchers have attempted to clearly define this concept within a given context. In the context of natural ecosystems, sustainability can be defined as the dynamic equilibrium between natural inputs and outputs, which are modified by external events, such as climate change and natural disasters (Fresco and Kroonenberg, 1992).

The concept of sustainability has had widespread influences despite the complexity of its inherent definition. The definition of sustainability is derived from many interdisciplinary viewpoints. Occasionally, this concept is categorized as strong and weak sustainability (Bell and Morse, 2008). Strong sustainability is related to ecological sustainability, with an emphasis on the environment. Here, quality is related to strong sustainability in terms of physical quantities, and weak sustainability is primarily related to financial parameters, which are conventionally based on cost-benefit analyses.

2.2. SCALES OF SUSTAINABILITY

Sustainability in urban spaces is highly dependent on surrounding areas. The outside world significantly influences urban regions, even when the boundary is well defined. Sustainability theory emphasizes the importance of spatial scale on sustainable practices.

Furthermore, scale is even more important when judging the sustainability level of an existing system. Problems may arise when implementing sustainability practices, even if the boundary is clearly defined. Thus, scale strongly influences sustainability. The proper temporal dimension is the time scale over which sustainability occurs. A starting point or reference condition is required to determine this development, and could affect the result. In consequence, spatial and temporal scales are basic components for determining the success of sustainability. Thus, could these two elements of sustainability be understood by defining the scales within which sustainability could be achieved? This definition may include spatial scales, such as system boundaries or political boundaries, as well as the time scale over which sustainability is to be achieved. These scales could be decisive when selecting a scale or reference point for verifying the results.

2.3. System Quality

In simple terms, system quality refers to an element directly related to the quality of life. Quality of life is a core element for elaborating the concept of 'sustainability' and 'sustainable'. Individuals constantly try to interpret their own spaces in life in terms of culture and values. All of their goals, exceptions, standards and concerns are linked to the culture and values of which they belong. A number of terms, such as 'well-being', are frequently used to explain the meaning of quality of life. However, some researchers consider these metrics as subjective. A simple inspection could identify how the important elements influencing the quality of life are related (Figure 2.1). The consumption of these elements indicates the quality of life and is related to the environmental quality and economic growth of a nation. Environmental quality is a quality-of-life indicator that consists of, factors such as environmental degradation, which is related to the per capita income according to the Environmental Kuznets Curve (EKC) (Figure 2.2). Environmental indicators at the regional or local scales, such as air pollution and per capita income, can be expressed using the linear, log-linear and log-quadratic forms of the EKC (Bartoszczuk et al., 2001).

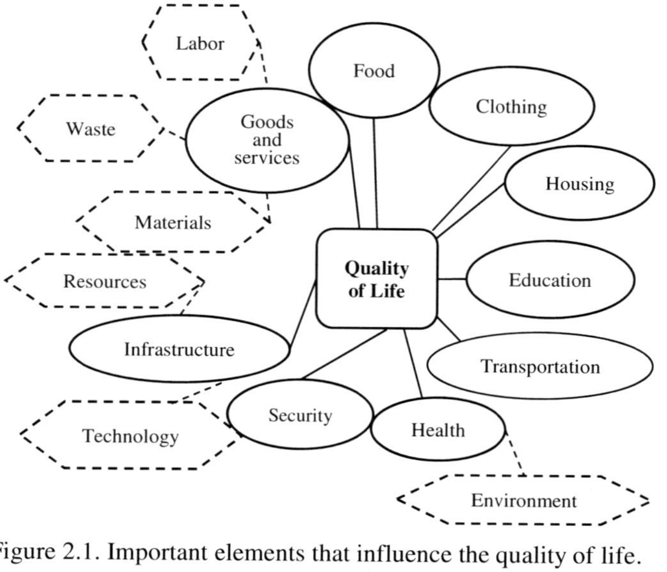

Figure 2.1. Important elements that influence the quality of life.

$$EI_{it} = \beta_0 + \beta_1 y_{it} + \varepsilon_{it} \quad (2\text{-}1)$$

$$EI_{it} = \beta_0 + \beta_1 y_{it} + \beta_2 y_{it}^2 + \varepsilon_{it} \quad (2\text{-}2)$$

$$EI_{it} = \beta_0 + \beta_1 \ln(y_{it}) + \varepsilon_{it} \quad (2\text{-}3)$$

$$EI_{it} = \beta_0 + \beta_1 \ln(y_{it}) + \beta_2 (\ln(y_{it}))^2 + \varepsilon_{it} \quad (2\text{-}4)$$

where *EI* is the environmental indicator, β is the regression coefficient to be estimated, y_{it} is the per capita income, ε is the error term and *t* is time.

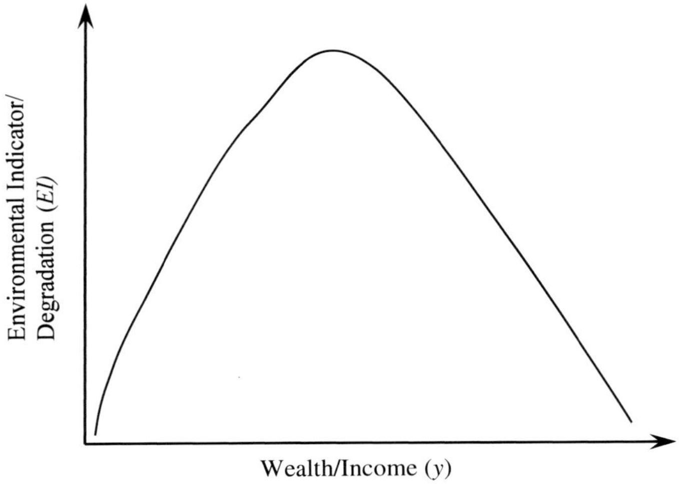

Figure 2.2. Environmental Kuznet Curve showing increasing degradation with income until a turning point.

The linear input variables can be written as follows:

$$EI_{it} = \beta_0 + \beta_1 y_{it} + \beta_2 y_{it}^2 + \beta_3 y_{it}^3 + \varepsilon_{it} \quad (2\text{-}5)$$

The first and second derivatives indicate the change of the curve, with values of zero at the absolute maxima. The first and second derivatives can be expressed as follows:

$$\frac{dEI_{it}}{dy} = \beta_1 + 2\beta_2 y + 3\beta_3 y^2 \quad (2\text{-}6)$$

$$\frac{d^2EI_{it}}{dy^2} = 2\beta_2 + 6\beta_3 y \quad (2\text{-}7)$$

Assuming that the coefficient β_3 is equal to zero, the above equation yields an inverted U-shaped curve with an absolute maxima if $\beta_2<0$. Again, $\beta_1 + 2\beta_2 y=0$, and the inflection point can be determined on the abscissa as follows:

$$y_{tp} = -\frac{\beta_1}{2\beta_2} \quad (2\text{-}8)$$

where y_{tp} is the inflection point, which is positive when $\beta_2<0$. Therefore, the linear input variables can be used to calculate the inflection point and establish the relationship between environmental indicators and per capita income in the EKC expressions.

2.4. Application of Sustainability

Sustainability must be applied to each parameter at the relevant scale to obtain the most effective output. Socio-economically appropriate and feasible methods must be used for long-term measurements. Well-defined metrics should be applied when observing improvements resulting in sustainable development (i.e., the fundamental elements, which are closely related to the goal of sustainable development). Thus, these elements must be addressed before any sustainability progress is made.

Recently, sustainable development has become a priority concern for international organizations, especially socio-economic aid agencies, development planners, researchers, and monitoring agencies. Sustainable development has become a universal theme at international conferences and in articles and journal publications, as reflected in mission statements related to environmental and social development. Although some critics with to examine the reasons for the popularity of this widely discussed topic, sustainable development is clearly an overriding theme today. Sustainable development is a model-specific case, which is quite different from what some see as a conflicting term of sustainable growth. Sustainable development has become a typical example of practical holism and simultaneously symbolizes an ultimate practicality because it would eventually be meaningless if it could not be achieved. The root of sustainable development lies in the present. If we can apply sustainability in a broad sense, it will enable us to compare and contrast all aspects of its application across multiple fields. In addition, we will be able to apply lessons from one area to another. Many specialists have mentioned the necessity of evaluating and measuring sustainability. This chapter elaborates a few approaches in this regard and discusses some problems that have been faced.

2.5. Roots of Sustainability

The concept of sustainability is derived from six strains of thought. These strains are separate but related to each other and have been used to discuss multiple issues, such as the rates of population growth, resource use, and pressure on the environment. Eventually, these six concepts, which are widely recognized as the roots of sustainability, were used to explain the concept of modern sustainability (Kidd, 1992).

1) Ecological/carrying-capacity: Carrying capacity is a view that states an ecological system can only sustain a limited population because every individual in that population consumes resources from that particular system. Thus, it is important to utilize resources in a sustainable manner. In this regard, Maximum Sustainable Yield (MSY) is a useful concept that is used to identify the maximum rate of resource utilization beyond which the entire system may collapse. Thus, carrying capacity is a core concept in ecology and can be found at the center of the other five strains of thought. 2) Critique of technology: Critique of

technology is a counter to the perceived indiscriminate use and exportation of technologies that may pose danger to the environment. When we talk of sustainability, agriculture is the top priority for obvious reasons. The end product of agriculture is food, which is a basic human need. On the other hand, land is the ultimate resource for agriculture, and the agricultural system needed to feed the world requires large areas of land. Hence, what occurs in the agricultural sector will significantly impact the environment and human society. Thus, one primary concern of the modern world is to make progress towards the development of a sustainable agricultural system and to promote such systems around the world. It is unclear if we can minimize or eliminate the technologies acting as barriers to sustainable agriculture or if minimization of such technologies is adequate. Thus, we must determine how much these technologies should be minimized. The answers to this question depend on the individual context and scale within which the individual has defined sustainability. 3) Resource Environment: Earth's ability to sustain a growing population. 4) No Growth or Slow Growth: Growth should be limited to avoid ecosystem collapse within the next few decades. 5) Eco-development: An approach for development aimed at harmonizing social and economic objectives with ecologically sound actions.

2.6. Ecosystem Indicators

Without oversimplifying a complex field, two broad approaches exist regarding environmental indicators that use the popular top-down approach and comprise two key methods. The first approach is to identify the 'indicator' species that are sensitive to changes in the environment, and the second approach is to gauge the state of the environment by measuring biodiversity. One example is the Shannon-Wiener index of biodiversity (H):

$$H = -\sum_{i=1}^{i=s}(p_i)\,(log_2 p_i) \tag{2-9}$$

where s is the number of species, and p_i is the proportion of the total sample belonging to the i_{th} species. The value of p_i is calculated as n_i/N, where n_i is the number of individuals in species i and N is the total sample size. The advantage of such an index is that it simplifies complexity into a single value that readily allows for comparisons, although it has several limitations.

2.7. Sustainability Indicators (SIs)

To identify a specific problem, a number of aspects associated with the problem should be considered. Next, these aspects are interpreted as indicators. Here, the values of the SIs of a system must be measured. The next step is to interpret the result and use the interpretations to solve the problem (Figure 2.3). However, it is difficult to determine how many indicators to use and which indicators are most relevant to the problem. Thus, one must know which component should be measured. The boundaries for sustainability are numerous and significant. The SIs are clustered according to the sustainability component they represent. Sustainability indicators can be divided into two groups, (1) state and (2) pressure. The state

SI describes the state of a variable. For example, a soil's physical and chemical properties can be used as an indicator to express the state of the environmental quality of an area. In contrast, the pressure SI measures a process that literally influences a state SI, for example, the rate of air pollution or the amount of chemical (e.g., pesticide) application that could impact the environment. Within this method of division, SIs are often clustered in different ways that depend on the aspect or scale of sustainability to be measured. However, a third type of SI, called 'response', has been included by the United Nations (UN) and is used to measure the necessary progress in the government's response. However, other organizations, including non-governmental organizations that work for the public on behalf of a government or social organization, would also be broadly counted in response SIs. These three category indicators are known as the PSR (Pressure-State-Response) framework (Figure 2.4). However, the UN realized that the pressure SI could only indicate negative effects in the SI measurement process, while adding another element, such as driving force, could indicate positive and negative effects. Thus, this framework could be further interpreted as a driving force-state-response (DSR) framework, which is similar to a cause-effect model in which driving forces affect the state and lead to a response. This framework is extremely popular and has been widely used in different contexts. To interpret various phenomena, this approach was further modified and is represented as DPSIR (driving force, pressure, state, impact and response). The impact of the SI is used to gauge the effect of a state, for example, the occurrence of child mortality in a country could result from the level of malnutrition. The DPSIR framework is represented in Figure 2.5.

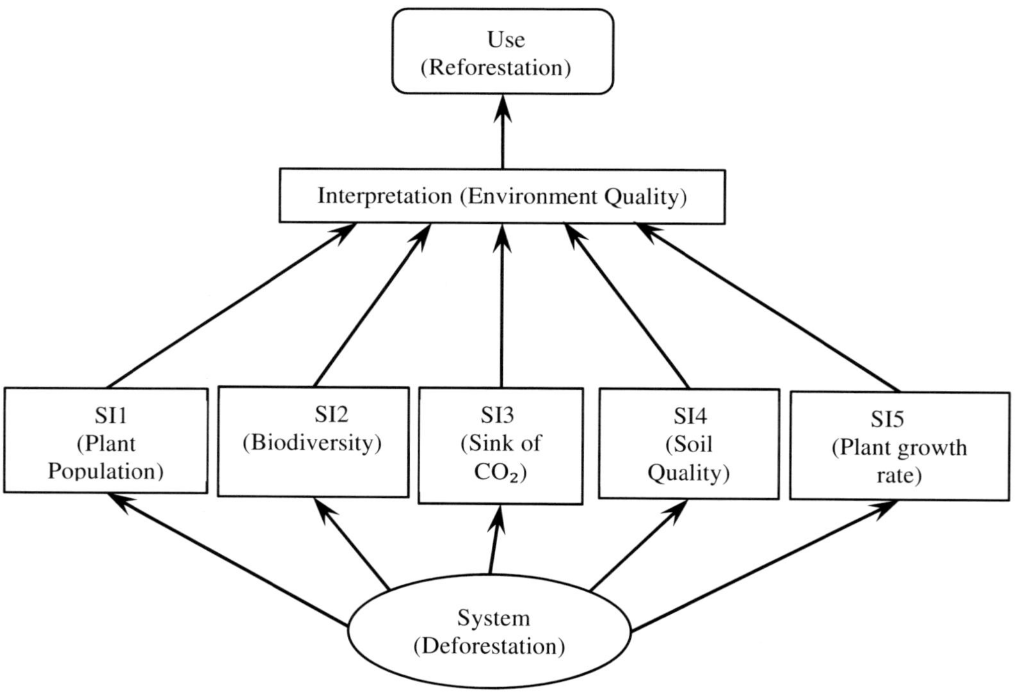

Figure 2.3. The basic theme of sustainability indicators (SIs) according to Bell and Morse, 2008.

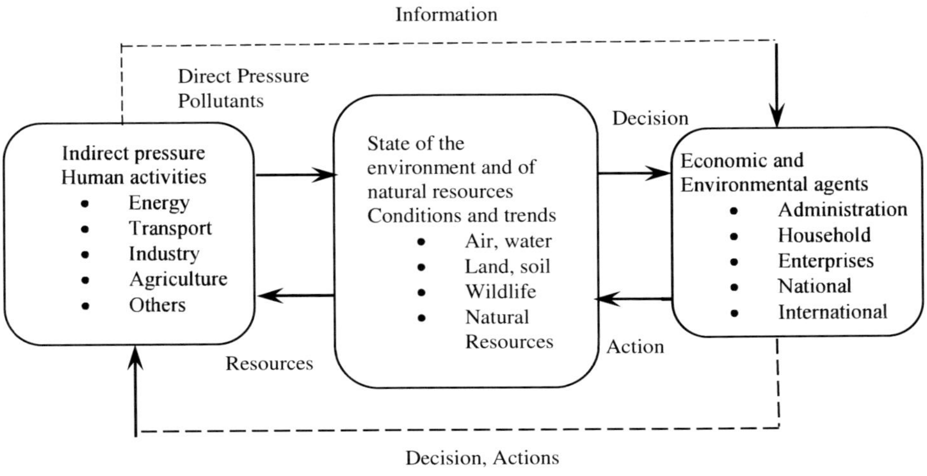

Figure 2.4. The pressure-state-response framework (OECD, 1993).

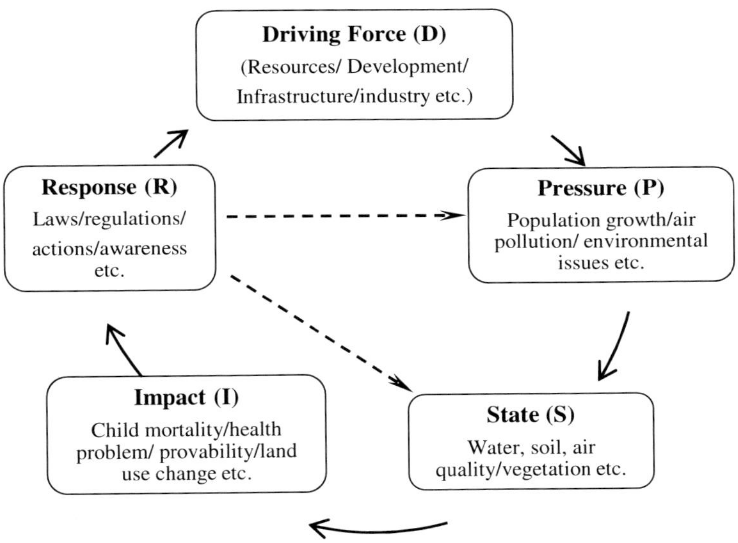

Figure 2.5. DPSIR framework for reporting environmental issues.

2.7.1. Environmental Sustainability Index (ESI)

One of the most striking examples of SIs is the Environmental Sustainability Index (ESI) that was developed by Yale and Columbia Universities in collaboration with the World Economic Forum and the Joint Research Center of the European Commission. The ESI is a composite index for tracking a diverse set of socioeconomic, environmental, and institutional indicators that characterize and influence environmental sustainability at the national scale. The ESI is presented as a single number, which is mathematically derived from a list of equally weighted indicators. The method used to derive the ESI is multifaceted. However, the related aspects comprise environmental systems, reducing environmental stresses, reducing vulnerability to environmental stresses, the societal and institutional capacities for responding to environmental challenges, and global stewardship (Figure 2.6). Overall, ESI is a benchmark that can be used to determine environmental performance, identify whether environmental results are above or below expectations, and note best practices and investigate interactions between environmental and economic performance. The ESI is calculated on a national state basis, and the value for each country is derived according to its state in regard to the mentioned aspects. Thus, this value differs from country to country and ranges from a minimum of 0 for the most unsustainable conditions to a maximum of 100 for the most sustainable conditions.

However, the most important concern regarding sustainability indicators is whether the indicator is an appropriate or feasible approach for measuring sustainability. Based on all discussions of sustainability, indicators for choosing the most suitable SIs from a subjective point of view and for measuring indicators appropriately are very challenging to identify. Sometimes the entire procedure, especially the selection of SIs, is subject to many pressures, such as agendas and biases. A number of criticisms of SIs, have been reported based on some major issues within the approach.

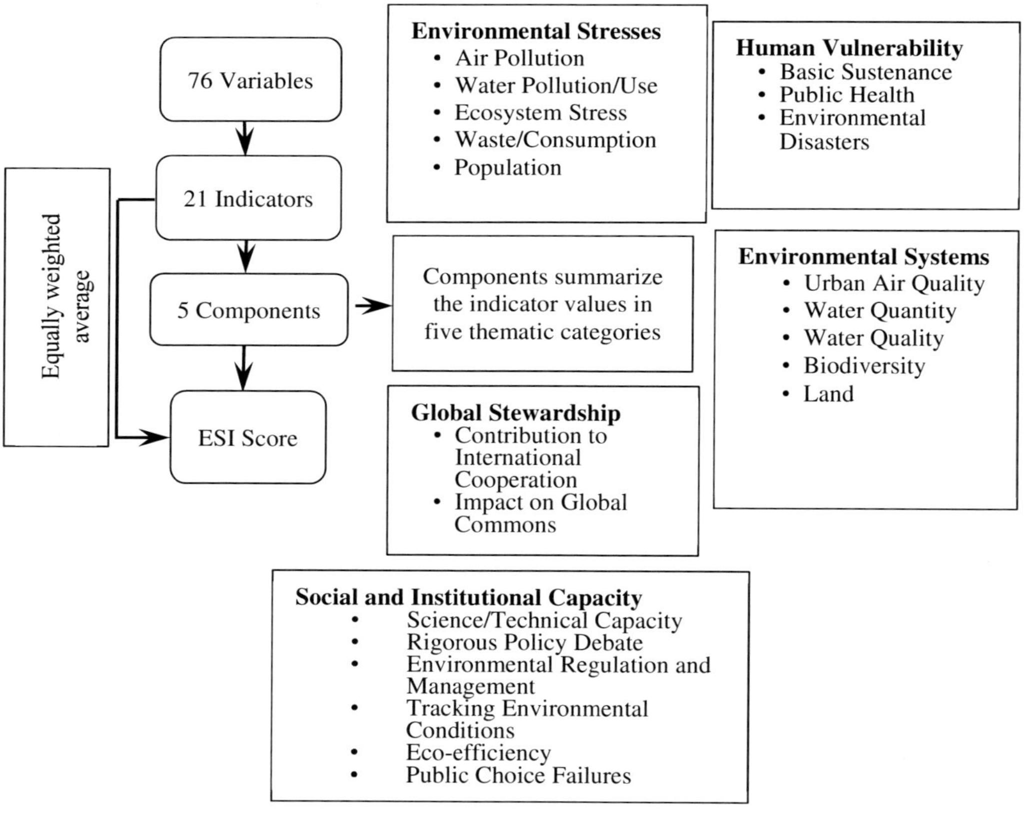

Figure 2.6. ESI frameworks with five components (Yale and Columbia University).

2.7.2. Maximum Sustainable Yield (MSY)

In this case, SIs can be examined using an important example of applying the MSY in the field of fishery management. Fishery management is the first time-dependent field that people attempted to use for optimizing resource management (Figure 2.7). This approach was based on fundamental ecological ideas derived from classic Western science. In addition, fishery management was subjected to the robust influences of ecology. For example, fisheries tend to have clearly defined spatial boundaries, including isotherms, physical boundaries, and areas of upwelling or divergence. Thus, as stated by the United Nations at the Rio Earth Summit in 1992, MSY is 'an expression of the state of fishery resource exploitation to its sustainable size'. Some even referred to the decade after 1945 as the 'golden age for the concept of maximum sustained yield' (Bell and Morse, 2008).

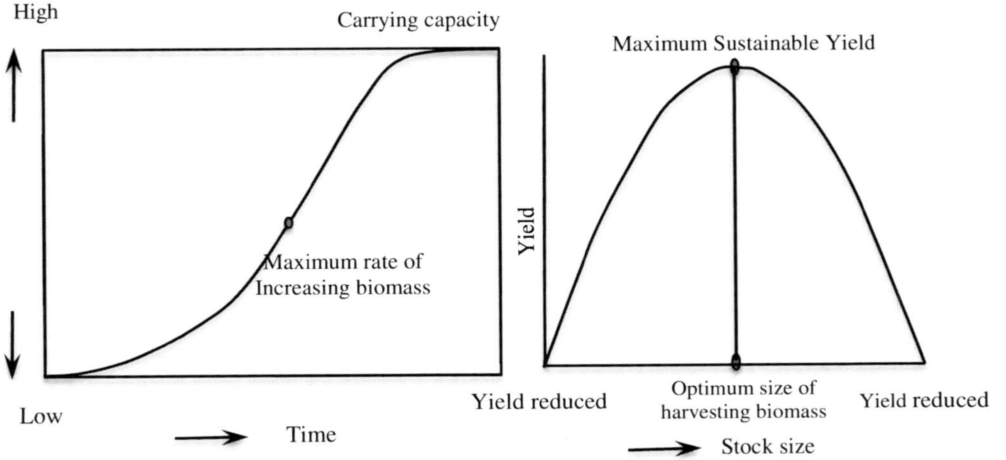

Figure 2.7. Concept of maximum sustainable yield for fishery management.

Although it is always very difficult to identify a spatial boundary for sustainability, fishery management has attempted to simplify this problem to the fundamental elements of population growth within a boundary. Many scientists throughout the historical literature have proposed various mathematical models for animal population growth. The central idea of these models is that population growth occurs through an accelerating process with respect to time. The basic concept used in population biology is derived from the first-order equation known as the logistics growth model, which was originally developed by Pierre-Francois Verhulst in 1983 and is often used in ecology to estimate the biological yield over a period of t for a population size of N.

$$\frac{dN}{dt} = rN(1 - \frac{N}{K}) \tag{2-10}$$

Here, r defines the growth rate and K is the carrying capacity. However, the following question arises: What are the conditions for carrying capacity for such population growth? We may consider the carrying capacity as constant, which is not necessarily true. As the environmental conditions vary, the carrying capacity may also vary. However, when the population reaches carrying capacity, the animals do not suddenly disappear. The population may exceed the carrying capacity for a certain time and then fall back below the carrying capacity limit. If the population exceeds the carrying capacity, the excess will persist for a certain period before the population disappears. When we discuss population, two groups must be considered, a group that takes part in population growth and a group that does not participate in population growth, including the very old and very young. The population growth model known as Leslie Matrices is used to predict the growth rate in proportion to the population (Bittinger et al., 2006). This model incorporates the proportions of individuals who survive and their reproduction rate, which could provide a better predictor of population growth. The proportion of the population that survives from one season to another is known as survivability. The number of offspring is called the fecundity of an organism, with the population projection matrix or Leslie matrix expressing the survivability and fecundity of a species. For example, consider a bird species with a certain number of hatchlings (h) and adults (a), written as $v = \begin{bmatrix} h \\ a \end{bmatrix}$. The Leslie matrix can be expressed as follows:

$$G = \begin{bmatrix} f_{afh} & f_{afa} \\ S_h & S_o \end{bmatrix} \tag{2-11}$$

where f_{afh} is the average fecundity of the hatchlings next year, f_{afa} is the average fecundity of the adults next year, S_h is the survivability of the hatchlings, and S_o is the survivability of the older birds. The next year's population can be projected by using the products of the following matrices:

$$Gv = \begin{bmatrix} f_{afh} & f_{afa} \\ S_h & S_o \end{bmatrix} \begin{bmatrix} h \\ a \end{bmatrix} \tag{2-12}$$

We can optimize sustainability by using population projections when animal resources are required for human livelihood. Additionally, some models are referred to as biomass dynamics models and act as a basis for some resource management models. These models build existing methods by including harvesting with the other components (births, deaths, immigration and emigration) of population growth. Thus, if immigration and emigration for a particular population remain constant, the number that can be removed by harvesting will only depend on the balance between births and deaths (the population growth rate). Therefore, the maximum difference between births and deaths is the maximum number of animals that can be removed without streamlining the population. This is referred to as the MSY. However, the concept of MSY is a recent development. It has taken a long time to apply this model, partly because data collection is required to establish the population growth rate, which requires very detailed knowledge about an animal population. The MSY concept has many variations according to the different definitions of 'sustainable' For example, in this context, Ecologically Sustainable Yield (ESY) is defined as 'the yield of an ecosystem can sustain without shifting to an undesirable state' (Zabel et al, 2003). However, the Maximum Economic Sustainable Yield (MEY) is related to the economic costs and benefits of harvesting and is used in fishery management. The MSY focuses on sustainability through population conservation, while ESY uses a much more comprehensive vision for sustaining some concept of a desirable ecological state and MEY takes an economic approach. The main purposes of these concepts are to ensure that MSY is a useful tool for sustainable resource utilization. However, the effectiveness and suitability of MSY must be demonstrated (Bell and Morse, 2008).

Surplus yield models (such as the Schaefer model) are used in population biology. However, these models are also subject to some limitations and should not be used as management models. Here, a problem arises when another surplus yield model is derived from a different starting point, such as when a model relates the current population size to the population size of the previous generation rather than time. Dempster (1975) showed that wild populations fluctuate widely (an effect caused by an array of environmental and biotic factors) and do not increase smoothly to a maximum before leveling off. Thus, the reliability of MSY will be weakened if a different unique starting point is used. As Pitcher and Hart (1982) stated, 'In its classical form, MSY excludes the effects of competition, symbiotic or commensal relationships with other species, tropic relationships or changes in carrying capacity due to pollution or other human influences.'

Factors that increase fish biomass include birth and immigration, and factors that decrease fish biomass include death, harvesting and emigration. If the "Immigration/Emigration" factor is positive, immigration dominates. However, if the "Immigration/Emigration" factor is negative, emigration dominates. Birth and death are influenced by predation from natural enemies. In this model, the following four loops are considered: the Reinforcing (R) loop associated with exponential increases/decreases; the Balancing (B) loop associated with reaching a plateau; the birth loop (R1) associated with an increase in fish biomass that results in increasing birth rates and results in greater fish biomass (Figure 2.8); and the death loop (B1) associated with greater death rates due to a decrease in fish biomass and with increasing fish biomass resulting in increased death, which decreases fish biomass. Another factor includes the following human effects on fish biomass (R2): harvesting increases lead to → a decrease in the shortage of food for human consumption, which → increases human population → and results in increased harvesting. If the human population increases, harvesting increases. Greater harvests reduce human food shortages and result in increasing human population. Finally, one must consider the effects of environmental destruction (B2). In this case, fish biomass increases → harvesting increases→ and the shortage of food for human consumption decreases. Greater human populations result in greater environmental destruction, which increases environmental degradation and results in decreasing immigration and fish biomass. From the R2 loop, harvesting increases the human population. However, increasing the human population results in environmental destruction. Environmental destruction is expected to result in the environmental degradation of fish habitats. Next, emigration will increase and fish biomass will decrease. Consequently, harvesting will decrease.

There is a risk in using such approaches as a basis for management models. For example, deriving another surplus yield model from a different starting point instead of relating population size to time is one problem that may be encountered. Therefore, by simply using another unique starting point, the foundations of the MSY appear to be insecure. However, the MSY has been popular for a long time because of its simple concept, the ease with which existing data can be inserted into MSY equations, and its ability to present sustainability in a single diagram.

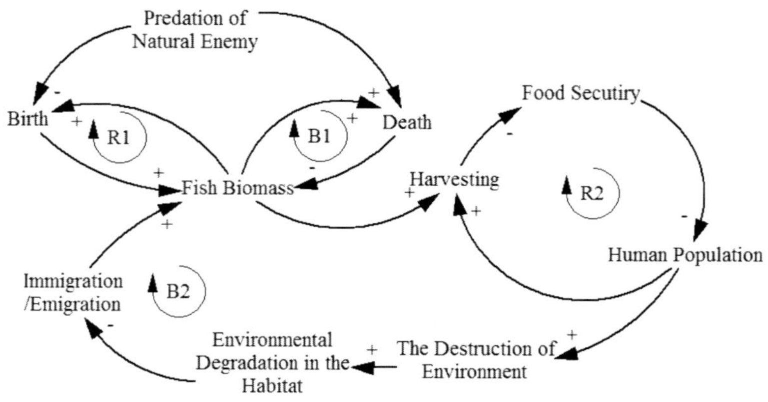

Figure 2.8. A system dynamic model for fish biomass change with different factors.

2.7.3. Sustainable Process Index

The Sustainable Process Index (SPI) is created using the operationalist form of the principle of sustainability. The SPI applies the operating data that are available at the initial level of planning and data regarding the natural concentrations of substances (not their presumable impacts, which are usually unknown). The focus of SPI measurements is the calculation of area, which is necessary for including a process that occurs completely in the biosphere (Narodoslawsky and Krotscheck, 2004). This area includes the area required for producing raw materials, processing energy and building facilities, as well as the area required for the staff and for the storage of products and by-products (Lundin, 2003).

2.7.4. Eco-Index Indicators

The European Outdoor Group developed the Eco-index for the outdoor industry as an assessment tool for designing sustainability practices (Chambers et al., 2000). This index frame work is modular and widely used for measuring the environmental footprint in the supply chain to identify areas for improvement and sourcing. In this method, the impacts of data related to the entire life cycle are used to define environmental footprint conversion factors, especially the possible key factors. The footprint is measured according to the global average yield, which must be normalized using equivalence factors.

2.7.5. Multi-Scale Sustainability Indicators

The most commonly developed multi-scale sustainability indicator is AMOEBA. Many researchers have introduced alternative approaches for overcoming the limitations of MSY. The AMOEBA approach provides a visual expression of a composite SI of the system, which is compiled from a different set of data regarding the various aspects of that system. Ten Brink et al., (1991) provided a comprehensive discussion of sustainability and stated that the three valuable characteristics of yield, biodiversity, and self-regulation should be addressed to express sustainability in a visual manner. Next, Ten Brink et al., noted that 'self-regulating ecosystems have low management cost' (i.e., there is no need to apply repetitive and costly human management if the system itself is strong). The next important feature of this approach is to assume that sustainability can be defined by how far the current ecosystem departs from an identified 'reference' ecosystem. The third key feature is to identify specific SIs and measure them to determine the 'key characteristics'. For example, it is common in ecology to measure certain species to specify pollution or disruption. However, concern arises regarding how the species should be measured. Thus, a list of criteria that must be met should be available for selecting these species. This list includes the following criteria: quantitative data regarding the abundance must be available for the species; the species must be susceptible to human interference; the species must be accessible for easy and accurate measurement; the species should have some 'indicative value' for the conditions of the system; and the species should have some 'political and social appeal'. We can understand the first four criteria because they refer to ideal ecological features (Ten Brink et al., 1991). However, the fifth criterion seems more likely to require a subjective approach. Thus, we believe that the fifth criterion may be more effective and appropriate for selecting species that society and authorities already know and understand. The last stage, which is the most expressive stage, is the presentation of data. Ten Brink et al., (1991) suggested that a graphical presentation of diversity would be more effective than calculated indices. Such presentations should be simple and transparent so that water authorities and policy makers can understand them easily.

One such simple visual presentation, known as an AMOEBA, is sketched in Figure 2.9. Here, the circle represents the 'reference' condition at a certain time, and the arms represent the condition at a subjective time, which is proportional to the reference conditions. The AMOEBA approach can be used to present the state of the ecosystem at different times, which helps to reflect changes in the system's sustainability with respect to time. In addition, this approach can help to predict the conditions of such systems if proper data and policy interventions are available. This approach has been adopted as a decision making tool in environmental management. However, the AMOEBA approach has also been criticized. First, this approach only represents the 'state' SI, not the pressure SIs, which results in concern regarding how the 'state' of a certain ecosystem could be managed without knowing the reason for that state. Another concern is the mechanism by which all indicators are accumulated in one simple diagram because separate indicators are weighted differently. In addition, concern also exists regarding the selection of the reference point. The diagram varies depending on the selection of a reference point, which is a core element of the AMOEBA approach. However, the AMEOBA approach proposed by Ten Brink et al. (1991) is a visual image of indicator species richness, which only applies to a small proportion of the species that exist in the ecosystem. The species richness could change with changes in the environment, which is difficult to model.

The AMOEBA approach can be used as a map-like tool for visual expression. The AMOEBA approach is used in projects to present a simple guide for the sustainability of project outcomes relative to the set of SIs for given outcomes. The shape of the AMOEBA could change depending on the situation and on the project team's decisions. However, the project outcomes can be considered sustainable when the AMOEBA approach approximates a perfect circle within the equilibrium band. One of the goals of AMOEBA is to represent SIs in a way that helps explain the sustainability process to stakeholders.

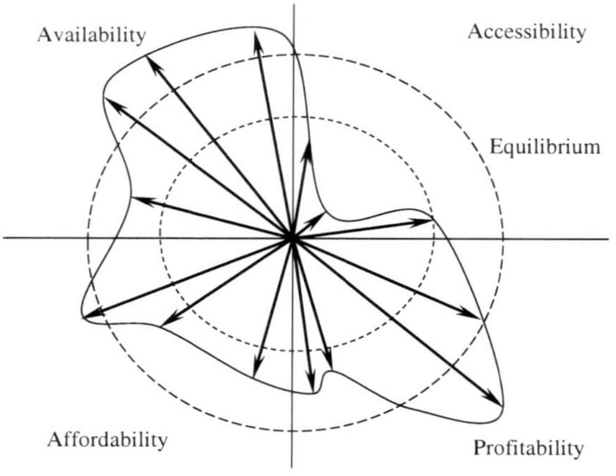

Figure 2.9. Representation of sustainability indicators in the local food matrix, AMOEBA.

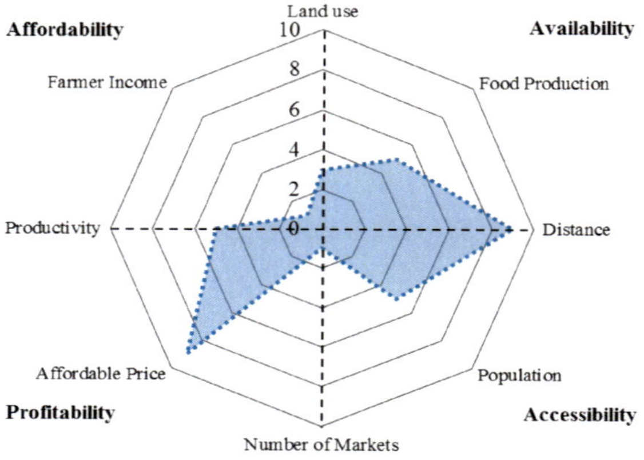

(a) 2011

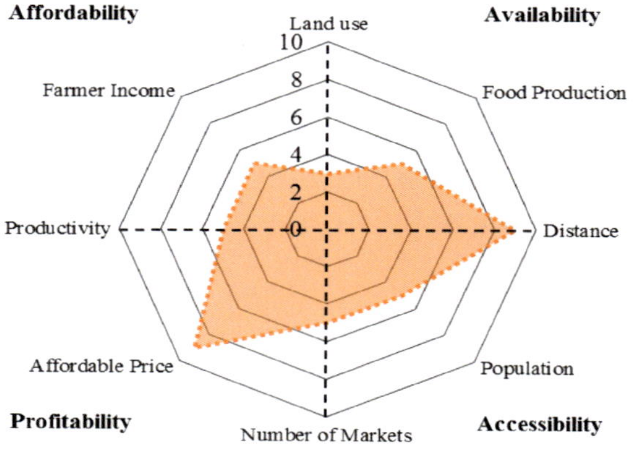

(b) 2014

Figure 2.10. Sustainability indicators in the local food matrix, AMOEBA using analytical hierarchy with nine scaling points (a) sustainability indicators in 2011 and (b) sustainability indicators in 2014.

This approach also encourages stakeholders to discuss the project's progress and accomplishments with the team. These sorts of visual approaches are helpful for systemic sustainability analyses (Bell and Morse, 2008). Such analyses can eventually help improve our understanding of the sustainability of certain aspects of a system or different aspects of different groups rather than influencing policy. Many examples discussed by Bell and Morse address multi-scale sustainability and the application of AMOEBA to ecosystems and the urban issues comprising social, environmental, ecological, and political indicators. Multiple sustainability indicators are also associated with urban issues in which quality of life is greatly influenced by urban population, migration to urban areas and resource availability. One example of local food production and its sustainability over time is illustrated using a 9 points scale and is based on the criteria and sub criteria obtained from a local district in Indonesia (Figure 2.10). The sustainability indicators were linked with policy implications in a visual expression to focus on long term planning and increase productivity and profitability.

2.8. Urban Sustainability

It is important to define sustainability in the context of urban communities. Urban sustainability is the process of developing a built environment that meets human needs while avoiding unacceptable social or environmental impacts (Hamilton et al., 2002). Humans wish to live in places where they can lead safe and peaceful lives with their family. These locations must include good surroundings with good neighbors, homes, local shops, adequate job opportunities, educational institutions for children, and hospitals. Dominski et al., (1992) stated that sustainability may be defined as a dynamic balance among three mutually interdependent elements, the protection and enhancement of the natural ecosystem, economic productivity, and the provision of social infrastructure, such as jobs, housing, education, medical care and cultural opportunities. With these visions, many organizations, consulting services, countries and communities are working to promote and sustain this new approach for sustainable communities. The Academy for Sustainable Communities (ASC) is one such organization that is working to provide the most efficient skills and knowledge for building sustainable communities nationally and internationally. The organization focuses on the seven following key elements that are needed to build a sustainable community: (1) Governance, (2) Transport and Connectivity, (3) Services, (4) Environmental, (5) Economy, (6) Housing and Built Environment, and (7) Social and Cultural. Sir John Egan suggested that sustainable communities must meet 'the diverse needs of present and future residents, their children and other users' by offering choices. To be sustainable, communities must ensure the effective use of natural resources, enhance the environment, promote cohesion and inclusion in society and strengthen economic prosperity. In addition, Egan introduced a concept widely known as the 'Egan Wheel' to justify the above-mentioned indicators for establishing sustainable communities (Figure 2.11).

The inhabitants of a community are the main individuals that can develop and implement sustainability practices within their boundaries. External specialists can assist them with solving critical issues or with maintaining the procedure. However, the members of the community have basic knowledge of their community and know the best ways to attain sustainability within their own context and limitations. Thus, the members become the ultimate specialists regarding their communities, and local knowledge gains priority over the period required for achieving sustainability goals. The people of a community can be regarded as a source of influence and strength for developing a community with a sustainable quality of life for present and future inhabitants.

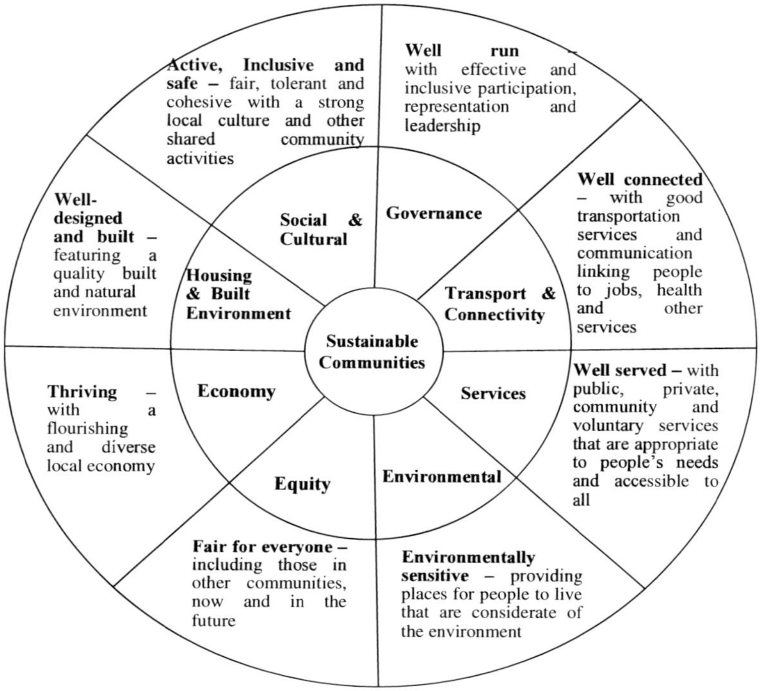

Figure 2.11. Egan Wheel for sustainable communities.

The people of a community may use distinct methods for ranking key components to ensure a sustainable community. For example, MacGillivary (1996) summarized the results of a survey that was conducted in a village in Scotland, in which the villagers were asked to rank the elements they considered essential for achieving a sustainable community. Another important example is the Norwich 21 set of SIs. In this set, SIs were placed into three categories to cover all of the important aspects that could be used to help achieve a sustainable community. The three categories are as follows: environmental protection, economic development, and social development. Community attitudes and systematic organization must be organized using proper institutional development. Therefore, institutional sustainability has received considerable attention in developed regions of the world.

2.9. Institutional Sustainability

Development requires a proper plan, and the execution of a plan must include proper initiative and implementation. Finally, to achieve the desired goal, an institution must evaluate the entire process. This institute could be part of a government or non-governmental organization or an international or local farm association that can help a country or community in sustainable development by producing proper plans and taking initiative for implementing and evaluating the process. Hence, sustainability in the context of institutions could be related to the financial independence of the institution and is also corresponds with the financers and the end beneficiaries. The role of the institutes has become more important for sustainable development and is a part of the core elements of the UN indicator set.

An institution can take the form of a government agency or ministry, an international agency, an aid agency or a non-governmental institution that plans, initiates, implements and evaluates development. Thus, sustainability in this context means financial self-reliance or some other measurable activity. Institutional sustainability has become a major consideration in the overall process of sustainable development and is featured within the UN indicator set. One reason for emphasizing institutional sustainability is that one significant activity of institutions is to provide financial services for the poor people of a community. Such financial support could allow these people to pursue sustainable development in different areas, regardless of available money. For example, the Garmeen Bank of Bangladesh has been pursuing these types of activities to allow rural people to become self-sufficient. In addition, SIs have been designed to evaluate the progress of institutional activities. Subsidy is one of the bottom indicators used to measure sustainability. The Subsidy Dependence Index (SDI) is an important index that measures the self-sufficiency of institutions that provide financial support to poor people, which allows individuals to sustain themselves and contribute to sustainable development (Bell and Morse, 2008). Another such indicator, the percentage of total costs covered by income, is also used. Nevertheless, no notable record of such financial service institutions achieving sustainability exists. In many cases, these institutions fail to perform due to lack of funding. However, some initiatives could be taken to encourage such financial service institutions, such as commercializing interest rates, covering operational costs and minimizing loan defaulting. By contrast, the supported institutions of some funding corporations must be separated into two units, one that provides the financial services and will eventually become sustainable and one that provides technical support and training and will rely on subsidies for future funding. The process towards active sustainability depends on several key issues, including the provision of financial and technical support to institutions.

2.10. Process toward Achieving Sustainability

This section focuses on the institutions that are concerned with maintaining sustainable development in their respective areas. Sustainable development is always related to a well-defined project within the context of a unique institution or branch of the institution. A project may be identified within a political or administrative boundary, such as a city, rural government area, a specific region of a country or a region of the world. Another property of the project is a definite time frame within which the project must meet its goal. Otherwise, the funding should include a scope to recommence work on the project depending on performance or another aspect. To an extent, the purpose of development activities may not only be limited to meeting the goal for the defined area within the fixed time but could also incorporate an improvement and monitoring process if the fund runs out.

Bell and Morse described a project as 'defined activities carried out by defined people with a defined end point in mind at a defined cost and over a defined period of time'. The most vital part of a project is funding, which depends on the availability of financial resources and may only be obtained from the stakeholders of the project's beneficiary. Thus, funders are always concerned about the progress and best outcome of a project to obtain the best value from the invested money. The money is awarded for a limited time and may only be used to meet the expected goals. However, the results must go through an evaluation and appraisal process to check whether the goals have been achieved. One indicator of such an achievement is likely to be financial self-sufficiency. For example, development projects may exist that could eventually become financially self-sufficient if they meet their goals and no longer need external funding to continue. However, to evaluate the diversity of project goals in different fields, such as education, health, transport, agriculture, etc., a range of project appraisal methods and techniques have been established and implemented that can classified into two comprehensive groups (Bell and Morse, 2008).

The first such technique involves a Cost-Benefit Analysis (CBA). The benefits of the project are analyzed and evaluated with respect to the project cost. A CBA requires that the benefits be expressed in monetary terms (i.e., it is fundamentally a quantitative method). Because this approach is based on economic values, it is also known as economic cost-benefit analysis (ECBA). However, CBA can also be used for social cost-benefit analysis (SCBA) by adding social values into the analysis. The second technique, Multi-Criteria Analysis (MCA), is a generic term covering a host of different approaches that may include an element of CBA. Some of these approaches, such as the CBA, may generate quantitative results, while others may be based on systems for scoring more qualitative effects (Munasinghe and Douglas, 2007). The MCA method is more flexible than CBA because an infinite number of criteria can be used in the MCA method. However, standardization is necessary because the result is highly dependent on the choice of the criteria and the weighting method used to relate these criteria to each other. Thus, MCA could be used as an evaluation method for SIs by replacing the criteria in the SI. We will further discuss AHP method of MCA in Chapter 5. To achieve sustainability, a clear mentality is needed, which is also characterized as a pragmatic approach by systems researchers. This approach has been developed in a number of ways.

2.11. Systems Thinking and Approaches

Systems thinking is a process of understanding how people, nature and situations partially or completely influence one another. In nature, systems thinking is evident in ecosystems in which various elements, such as air, water, land, plants, and animals interact with one another. Similarly, organizational management and research involves systems thinking. Logistics systems consist of people, structures, and processes that work together to benefit an organization or allow it to be sustainable. The end product of an effective management system is synergy, which indicates that the end product equates to a larger value than the sum of its individual parts. Generally, systems are composed of multiple components when designing or resolving a problem. These components include inputs (manpower, time, energy, information); processes (methods, software, analyses); outputs (products, reports, plans); and feedback mechanisms (information, reviews). In addition, some alternative systemic methods consider and solve problems that cover a diversity of subjects and include variations in thinking and limits in scientific methodology. Such methods lead to a systemic approach to problem solving and introduce a range of system approaches, new definitions, new thinking, and holism, eclecticism, systematism, and emerging premises for SI development.

Deviation of individual opinions tends to create changes in thinking, which is referred to as a paradigm shift. In other words, the movement of one paradigm (a set of assumptions about the world) to another is called a paradigm shift. Bell (1996a) illustrates this shifting mind-set in the context of a continuum. A spectrum or continuum describes a range of thinking from a single viewpoint, which can be used in any problem-solving case. This range varies from the most reductionist to the most holistic cases. According to Bell (1996b), a reductionist approach rejects ideas of reality and the importance of unscientific aspects of life, such as hunches, guess-work, instinct for rightness and even, in certain circumstances, illogical activity (i.e., activity that is inconsistent with narrow definitions of efficiency). Through empiricism, the universe is observed as fixed, knowable, measurable and predictable. In contrast, Webster (1995) indicated that holism is a theory that states that the universe, and especially living nature, is correctly observed in terms of interacting wholes (such as living organisms) that are greater than the sum of the elementary particles. In practice, holism always involves scientific and reductionist approaches of thought. By definition, holism is not holistic if it is viewed as exclusive or extreme.

Here, the main purpose of system approaches is for problem-solving. Thus, we must elaborate regarding the word 'system' as considered from different perspective. For example, a carbon cycle is defined as a physical system, a political constitution is defined as a social system, and the idea of sustainability is an abstract idea with some defined boundary. A relationship could exist between a system and its environment, but the extent that the environment affects the system will highly depend on the system itself. Many systems-based approaches are available for problem solving, each with its own criteria and perceptions. A few of these important approaches have been discussed in detail by Bell and Morse, and a summary is provided in the following section.

2.11.1. Soft System Method (SSM)

In this section, soft is used to describe a situation in which a problem's solution is not accessible at the beginning stage and when the objectives are not clear, which results in mixed purposes. Elements for SSM systems are defined comprehensively. Because we assume that problems may not be clear or structured, the system first states that considerable time should be allowed for observing the problem and recognizing inherent tasks and related issues. The next step is to determine whether the root definition of all related systems could directly or indirectly affect the case because the solution could also be obtained from other fields. After identifying and analyzing the root cause, an activity plan is developed that is based on feasible conceptual models. Then, the most important step is to compare this stage with the initial view of the problem. Omitting this comparison would allow the analysis to proceed under false assumptions. Next, a discussion of the analysis with all stakeholders should be conducted and each stakeholder must perform an individual analysis from their own field and perception. However, the combination of analysis and discussion before pursuing a solution is the strongest point of this approach. At the end of this stage, the process is repeated for the next issue. In this way, the process attempts to find approaches that can be followed by using a mutual understanding among all responsible stakeholders rather than by fixing the problem directly. Thus, the most important aspects of the soft systems approach result in a systemic way of thinking about the problem and in the involvement of all stakeholders.

2.11.2. The Learning Organization Approach (LO)

This approach was presented by Senge et al., (1994), who proposed that the following five disciplines could help an organization become a perfect model: systems thinking, personal mastery, mental models, shared vision, and team learning. These practices encourage the development and sustainability of the organization. The core objectives of this approach are group involvement, learning, and unanimity in decision making, which are more important than the independence of reductionist thinking. This approach has been practiced by many organizations and academic institutions regarding various issues and has been used to develop methods for meeting their goals. Thus, the learning organization approach eventually resulted in numerous high-range approaches and implementations. The basic difference between the learning organization approach and the SSM approach is that the main features of the LO approach include the behavior of the individual systems that facilitate learning. However, in the SSM approach, all sequential processes are important and systematic analysis is related to cycles of understanding.

2.11.3. The Participatory Rural Appraisal Approach (PRA)

This approach is the most important method from a socio-economic perspective and can be used to solve the issues created from project assessment (which is more likely to focus on the cost benefit issues of a project). The PRA approach has been widely applied and developed globally for various issues using numerous techniques. Chambers (1997), the main author of this approach, specified the three following pillars of PRA: (1) the behavior and attitudes of the development professional; (2) the need for sharing between different actors; and (3) a requirement for participatory methods. The SSM approach has many supporters, and the LO approach has been adopted by many management science experts. In addition, the PRA approach has been used by development organizations as a tool in the project procedures. The PRA approach helps to identify the stakeholder population involved in a project and highly emphasizes the perceptions of this population. Overall, the SSM, LO and PRA approaches all incorporate the stakeholders involved in particular issues and include a process of understanding, responsibility for the sustainability of the organization and an appropriate place for developing the analysis.

2.11.4. The Logical Framework Approach (LF)

This approach is the most relevant approach for project planning and management and inspires users to precisely and widely consider their projects and comprehensively represent them to stakeholders. Including a participatory element in this approach could enrich the process. However, the success of this approach is highly dependent on the commitment of the project team to work efficiently and effectively. The LF approach can be descriptive and analytical (Figure 2.12). The example has been provided for local food production. If it is descriptive, the log frame of a project is presented in soft and hard or formal and informal terms. This method allows a team of stakeholders related to a project to fix the core aspects of the project, by developing activities to attain the expected outputs and purposes and achieve objectives. This is the formal or hard element of the LF approach. In addition, this approach allows participants to establish assumptions for each stage, which are referred to as the informal or soft elements of the approach and are mostly placed in the column to the right of the logframe. The middle two columns focus on monitoring and project analysis and present the performance of the project in qualitative or quantitative terms. Thus, performance can be measured at each stage. For example, the performance of activities is judged by the amount of money spent and the achievement of activities to date. As an output, an overview of the project's impacts is presented, which reflects whether the project has achieved its initial objective. Finally, at the purpose level, the evaluation checks for the expected result. Overall, the approach is a systemic process that is coordinated at each level but maintains unity as a whole. This approach typically involves a range of participants in all processes of the project and has the flexibility to modify the project to adapt to a changing environment or situation while still relying on control and proper guidelines.

Table 2.1. Logical framework for developing a logistics model for a local food system.

Narrative Summary	Objectively Verifiable Indicators (OVI)	Means of Verification (MOV)	Assumptions
• **Goal** • Food Security • Sustainable Agriculture	• Increasing the competitiveness index • Reducing Imports of rice and wheat	• Report by World Economic Forum • Statistics and Database	• Support from FAO
• **Purpose** • Connect local farmers (producers) to the consumers • Shorter supply chain	• Improve economic viability • Increase farmers' income • More healthy food	• GDP • HACCP	• Support from FAO • Support from Government • Food System Policy from Policymaker
• **Outputs** • System modeling of food • Spatial data of local food distribution • Policy of food distribution	• Certification of local food • Local food guide map • One province one product	• Validation system of local food trace • Using GIS system • Monitoring and Evaluation	• Support from Government • Food System Policy from Policymaker
• **Activities** • Food movement • Map Development • Software Development	• Local food campaign • Spatial system GIS • Android /iPhone development	• Participation of more than 5000 people • GIS analysis • Verification of new software of local food guide	• Support from Government • Food System Policy from Policymaker • Student college participation • Farmers and Villagers participation

Sustainability is a concept that depends on multiplicity in perception and individual context and is regarded as a fundamental approach for development and achieving goals. The survival of future generations is highly dependent on the current use of the world's limited resources. Thus, sustainability plays an important role in every aspect of life. The highlighted approaches in this chapter present possible methods for practicing sustainability. Because the core issues of human life are to secure adequate food for everyone, agriculture has become a modern target for sustainability. Sustainability indicators, especially ESI and MSY, have received significant attention due to their impacts and the necessity for present practices. Thus, we discussed system quality and EKC to provide an overview for understanding environmental degradation. The bottom line is the focus on sustaining biodiversity, environmental quality, agricultural sustainability, and energy security for future generations. Throughout the world, fossil fuels are the main source of energy, and fossil fuel depletion and global warming are topics of concern for many nations. Continuing population growth and changing lifestyles will require food and energy security in the near future. Bioenergy campaigns and increased bioenergy potential have been discussed during the last fifteen years. A number of first, second, and third generation bioenergy sources have been reported. On one hand, agricultural production must be secure and energy security must be addressed. Regarding energy security, biofuel crops have been recently debated and ethical issues have been raised in the context of meeting demands for food and energy security. To meet the demand for food, intense agricultural practices are needed because their influences on soil quality and water confront health and environmental issues for environmental and agriculture sustainability. Again, to meet the demands for energy, bioenergy crop cultivation is intensively practiced, which raises questions regarding land use change policies. In the next chapter, we discuss how agricultural sustainability affects some aspects of agriculture, such as soil health, water quality, biodiversity, and the education of farmers. The intergenerational ethical concerns of biofuel production and environmental recovery are also be addressed.

Chapter 3

SUSTAINABILITY AND ETHICS IN AGRICULTURAL PRODUCTION: DEBATE OF FOOD SECURITY, BIOFUELS AND THE INTERGENERATIONAL ETHICS OF BIOFUEL PRODUCTION

In recent discussions of the sustainability forum, continuing debates over ethical issues in agriculture, food production, and the environment have gained attention. These debates are based on food security due to biofuels. This chapter discusses key issues for soil and water quality in agricultural production and the confrontation between food and fuel, such as bioenergy and crop cultivation. At the end of this chapter, intergenerational ethics are discussed with the continuation of trade-off solutions between the generational biofuels and the implications for environmental recovery.

Worldwide, the terms 'agriculture' and 'agricultural system' have been used in various aspects in the production of plant and fodder materials for animals, fibre and other uses. Agriculture is a complex process that involves biological, physical, socio-economic, and political constraints, which control sustainability in agricultural policy planning (Yunlong and Smit, 1994). Sustainability becomes a popular word that explains continuous systems of environmental policy and research. Some terms, such as, "Sustainable Development", "Sustained Use of the Biosphere", and "Ecological Sustainability", are widely used. The concept of sustainable agriculture describes the characteristics of the long-term maintenance of natural systems that request optimal production with minimum input, adequate income, fulfillment of basic food needs, and the demands of rural communities (Brown et al., 1987). All definitions of sustainable agriculture combine environmental, economic, and social harmony in an effort to realize the meaning of sustainability.

Sustainability is a concept that cannot be measured directly because appropriate indicators must be selected to determine the levels and duration of sustainability. The indicator of sustainability is a variable that is used to describe and monitor the processes, states and tendencies of systems at the farm, regional, national and worldwide levels. An indicator of sustainability must consider temporal and spatial changes and be predictable, measurable and interactive. The most recent issue today is how to design appropriate technologies that are compatible with social, economic, and ecological viewpoints. The index of biophysical sustainability, soil and water conservation, fertilizer use efficiency, energy use efficiency, and the productive permanence of the forest are suitable indicators for assessing natural resources (Hayati et al., 2010). Appropriate indicators include measurable and quantifiable environmental statistics that measure or reflect the environmental status. The concept of indicator was extended to the system statistics that measure national agricultural systems. Selected indicators of the criteria that are used to discriminate between factors can be causes or effects involved in controlling a system. The definition of criteria may be broadened to standards or rules that govern system condition judgments rather than single environmental conditions. The FAO introduced the tools used for agricultural systems, including cropping, farming, sub- regional, regional, and national systems during the 90's.

3.1. Agriculture and Food Production System

When an agricultural system is sustainable, the food security of the population is related to producing enough food. However, hunger and malnutrition are not only problems of food availability but also of food access (Kleemann, 2013). Hunger occurs when people cannot access the amount of food that they need. Improvements in food security and in the local agricultural sector are important for increasing the productivity of farming system. It is effective to meet the dietary needs of the rural population, which strongly influences other sectors of the economy. Both policy makers and analysts need to improve agricultural productivity and food security to achieve Millennium Development Goals (MDGs). For example, from the African perspective, fighting hunger is the most complicated social challenge. In many parts of Sub-Sahara Africa, the land is already strongly degraded. Thus, the methods for improving productivity by using mineral fertilizers and pesticides have damaged the foundations of the land for long-term productivity. The global demand for grain will increase by 75% between 2000 and 2050 (IAASTD, 2008). Thus, decreasing amounts of water and soil must feed more people. Moreover for analyzing agricultural systems concerned with environmental conditions, especially soil health and wastewater quality, the points discussed below must be considered.

3.1.1. Soil Health

In the sustainability concept, soil health indicates the capacity of a soil to be used productively without producing negative environmental effects. Soil health, emphasizes the integration of biological components with chemical and physical measures of soil quality (Figure 3.1). Soil health is very important to farmers and growers because they need to understand the sustainability of yields and the long-term impacts of soil management on environmental quality. However, the different aspects of the soil are interesting and each aspect cannot be observed independently. To maintain good soil, all of the unique aspects of the soil physical, biological, and chemical processes must be balanced and good. Soils can supply adequate amounts of nutrients, if their pH values, moisture contents, aeration, root penetration, and biological components are balanced. Soil health is influenced by inherent soil factors, including soil texture, fertility, compaction, and erosion.

Soil health assessments depend on the measurements of carefully chosen soil indicators to determine the soil health or quality. Sensitivity management is necessary for demonstrating distinct changes under contrasting practices. However, the variability and the measured values of the variable factors make it difficult to show distinct difference between management practices. The relevant measurements of soil processes include the measurements that integrate functional soil processes, such as infiltration, aeration, and nutrient cycling.

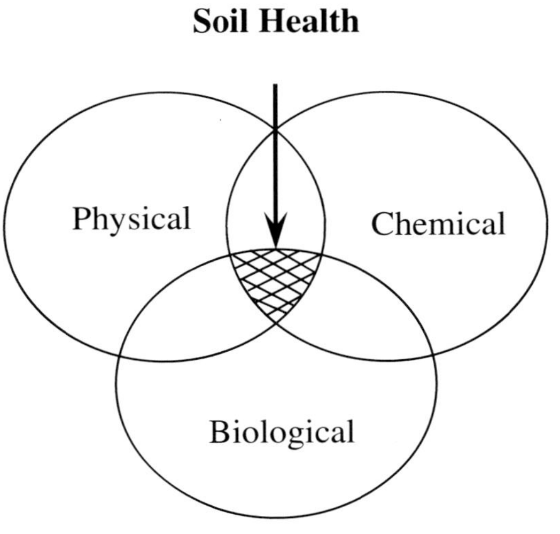

Figure 3.1. Soil classifications.

3.1.2. Wastewater Quality

Wastewater quality is an indicator that is used to evaluate the suitability of wastewater for disposal or re-use. Examination is required for selecting the results for intended use or discharge locations, including the physical, chemical, and biological characteristics of the wastewater. Water temperature is one of the most important characteristics of an aquatic system and is affected by dissolved oxygen levels. When the water temperature increases, the solubility of oxygen decreases. The chemical processes affected by temperature also influence the solubilities and reaction rates of chemicals. The rate of chemical reactions increases the water temperature (RAMP, 2014). In contrast, biological processes affect temperature and influence metabolism, growth, and reproduction.

3.1.3. Biodiversity

Biodiversity indicators are statistical measures of biodiversity that can help scientists, managers, and politicians to understand the conditions of biodiversity and the factors that affect it. The UK Biodiversity Indicators were developed and published between 2007 and 2010 (Figure 3.2).

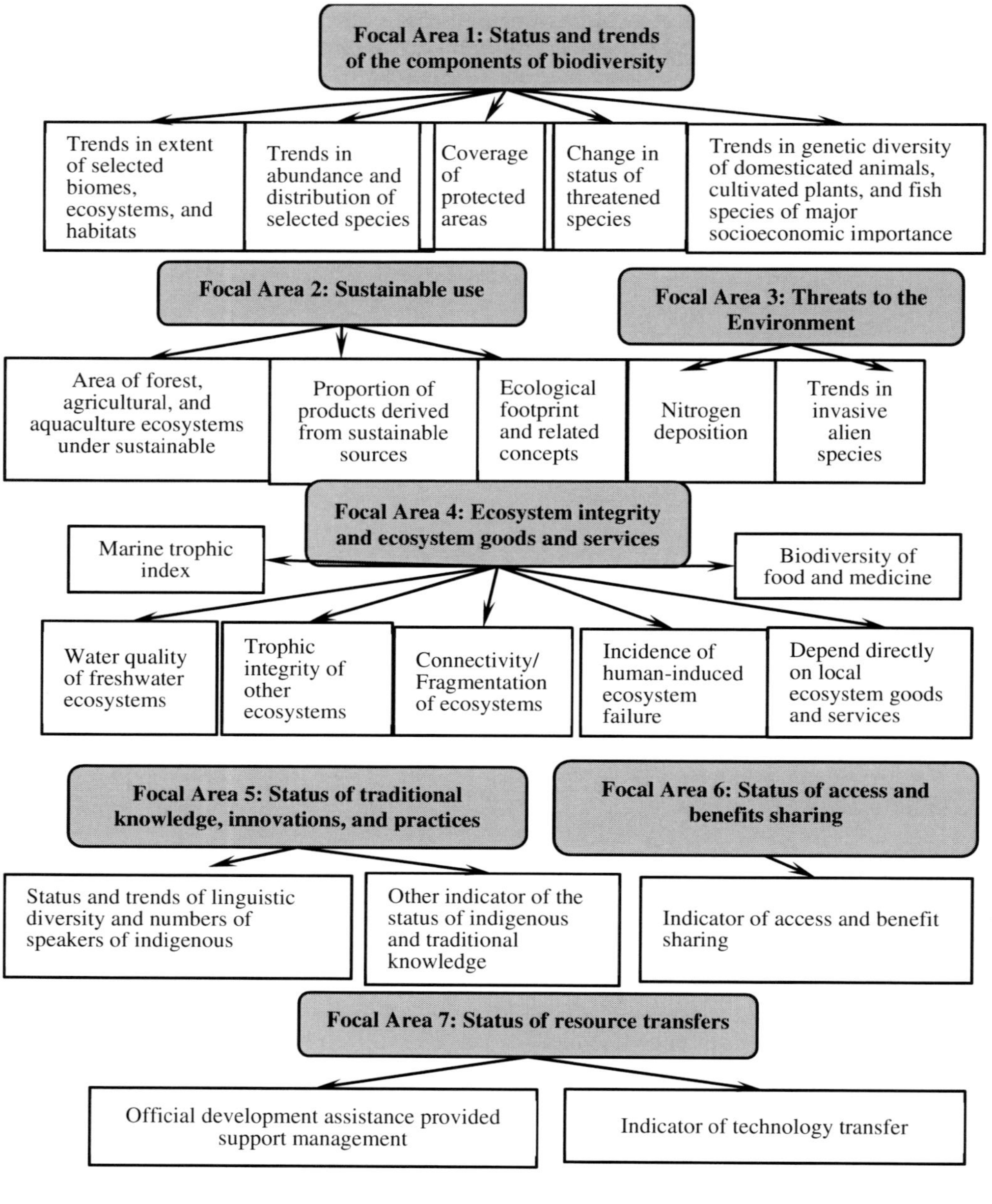

Figure 3.2. Biodiversity indicators (Georgina and Jonathan, 2007).

The UK has the opportunity to obtain large amounts of information regarding biodiversity that has been collected for a broad range of species and across a broad range of habitats by professionals and expert amateurs. This information provides an essential source of evidence for reporting changes in biodiversity and the impacts of policies and actions that preserve biodiversity. Indicators are one method that the UK can use to communicate monitoring and surveillance results. The audience for indicators is extremely broad, and ranges from general to public audiences in all portions of the private and public sectors.

3.1.4. Farm worker Education and Training

Sustainability is a concept that cannot be evaluated directly. Sustainable agriculture promotes environmental, economic, and social harmony to attain the meaning of sustainability. Appropriate indicators should be chosen to measure the levels and duration of sustainability. In this report, the indicator was extended to a statistics system that is used to evaluate changing conditions in an agricultural system. The considered system depends on multiple factors, such as soil health, wastewater quality, biodiversity, and the education and training of farm workers, which are used in our sustainability index for agriculture (Craig and Perrow, 1977). Every farm worker should know certain information, including what symptoms of illness are hazards to food safety. These symptoms include diarrhea, fever, vomiting, jaundice, sore throat with a fever, or an abscess full of pus on exposed skin. When these symptoms are present, farm workers must alert their supervisor. As part of training, growers should indicate the locations of first aid kits and outline policies regarding injuries, blood and bodily fluids in the field. Growers are required to state that all injuries must be recorded by reporting them to a staff member responsible for recording injuries. The firm's policy on the contamination of product in the field with bodily fluids should be fully explained in addition how to handle affected areas.

3.2. Food vs. Biofuel Production

Some experts argue that biofuels provide a solution for reducing the impacts of climate change and global warming because they are considered to have zero emissions. Meanwhile, other experts argue that biofuels may increase the risks of food shortages, especially for populations in developing states. Furthermore, risk trade-offs can be made to reduce certain risks, such as global warming, that actually increase the risk of food security. Several environmental risks that are related to biofuel issues include the depletion of fossil fuel production, a reduced dependency on fossil fuels, a reduction in GHG/CO_2 emissions, food security, changes in habitat and ecosystems due to the conversion of land for energy plantations, social issues and the economic growth of local inhabitants. The production of biofuels vs. food security represents one dilemma regarding the risk of diverting farmlands or crops to biofuel production on the food supply. Increasing use of farmlands or crops for biofuel production could result in higher food prices and price volatility, which would result in hunger.

3.3. World Food Production and Population

The world's population is predicted to reach 9 billion by 2050, which is nearly 6.8 billion more than the current population and would result in a substantially greater food demand (Figure 3.3). The increasing population must correspond with an increase in food supply. The World Bank estimated that cereal production must increase by 50% and meat production must increase by 85% between 2000 and 2030 to meet growing demands for food. In early 2009, the food crises persisted in 32 of the 36 countries affected in 2008. Indeed, global food reserves are currently lower than they have been in the past 30 years. Thus, the world needs to grow more food and include good policies for biofuels to achieve sustainability.

3.4. Biofuel Production in the World

In 2012, the FAO projected an average worldwide biofuel production growth rate of up to 7% a year. Global ethanol production is projected to increase rapidly, reaching 127 billion liters in 2017. This production will mainly occur in the United States, Brazil and (to a lesser extent) the European Union and China. Global biodiesel production is expected to reach 24 billion liters in 2017. Production trends indicate that ethanol and biodiesel supplies are steadily increasing; however, the global ethanol market is more than four times larger than the global biodiesel market. Both of these markets are increasing in traditional established markets, such as the European Union, Brazil, and the United States, and in other countries, such as China, India, and Argentina (UNEP, 2011). The latter countries are beginning to understand the economic potential of the biofuel sector and its role in the green economy. Despite increasing markets for biofuels, the global bioenergy potential is largely underused, particularly in regions with significant potential for efficiency gains in agricultural production and biofuel conversion.

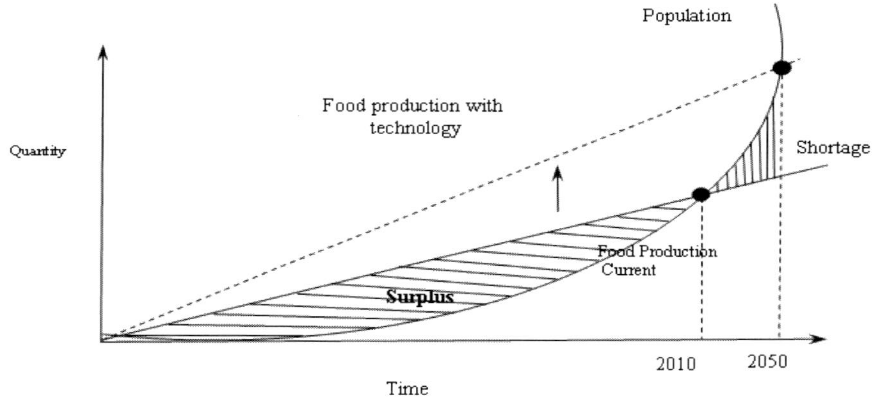

Figure 3.3. Current vs. future food production relative to the worlds growing population and a hunger map (United Nations World Food Program 2009).

By 2030, biofuels should meet 4% of the fuel demands for road-transport worldwide. Biofuel use has increased by 1% since 2012. More than one-third of maize production in the United States in 2008 was used to produce ethanol, and approximately half of the vegetable oil produced in the European Union was used for biodiesel fuel. Globally, irrigation water allocated to biofuel production is estimated to equal 44 km^2, or 2% of all irrigation water. Under current production conditions it takes an average of roughly 2,500 liters of water (about 820 liters of it irrigation water) to produce 1 liter of liquid biofuel (UN, 2009). For example, rain fed *Jatropha* is used as a biofuel and requires less water than many species and results in lower biofuel outputs. In this case, different water footprint measurements are obtained. As shown in Figure 3.4, sugarcane is one example of how these figures could be confusing because sugarcane is a water-intensive crop and, depending on local conditions, can have a lower water footprint relative to fuel output. In some countries where the biofuel industry is established, the first feedstock utilized is generally drawn from among the most important crops in the country. The most common examples of this situation include the use of corn in the USA, rapeseed in the EU, sugar cane in Brazil and oil palm in Malaysia and Indonesia. Biofuels tend to be led by a few dominant crops that are targeted by active policy support programs that account for domestic biofuel consumption patterns (Elbehri et al. 2013).

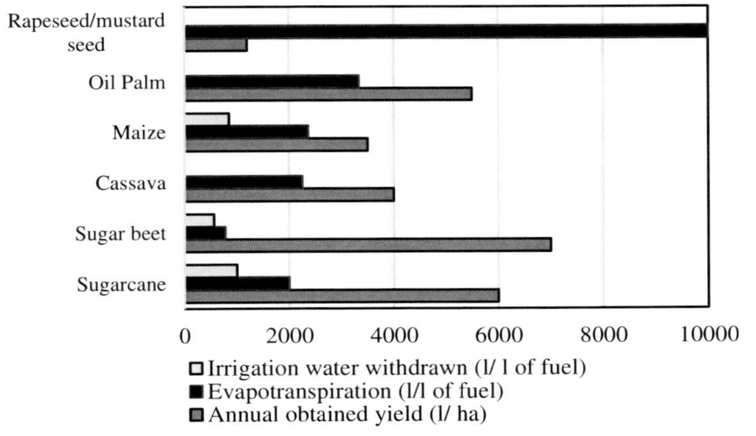

Figure 3.4. Water required for biofuel crop production (Fuel products and Energy densities: biodiesel 35 MJ/l, ethanol 20 MJ/l; ethanol (sugar): sugarcane, sugar beet, cassava, maize; biodiesel: oil palm, rapeseed/mustard, soybean) (World Water Assessment Programme, 2009).

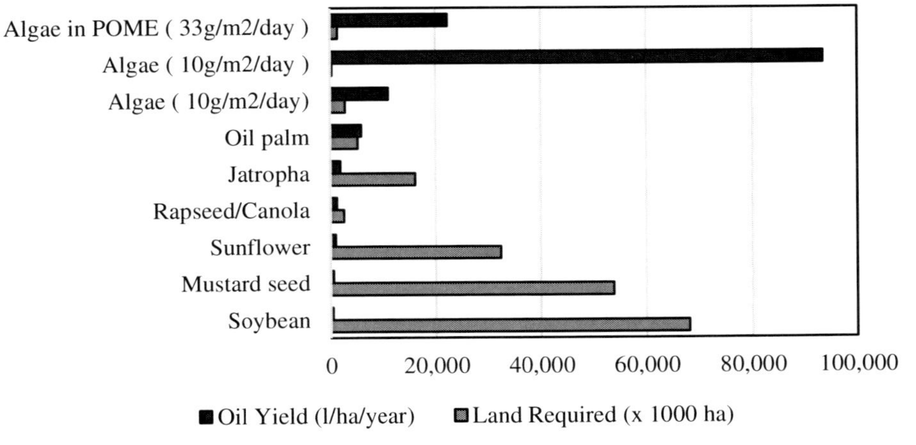

Figure 3.5. Estimated land requirement for biodiesel crops in 2025 in Indonesia for each specific crop and algae based on 30% lipids (Chisti et al., 2007 and Nugroho and Noguchi, 2014).

Another important motivation for biofuel use is energy substitution for including fossil fuels. Oil production varies greatly among different biofuel feed stocks and land requirements (Figure 3.5).

3.5. World Food Prices

The application of biofuels affects food security. For example, the use of biofuels will result in crop shortages and increase food commodity prices. This issue has been discussed by various studies in which biofuels are argued to be responsible for increasing food prices in recent years. The FAO/OECD concluded that food prices have been extraordinary volatile since 2008. In addition, they noted that biofuel production aggravates the price inelasticity of demand. It cannot be denied that the high demand for biofuels contributed to an increase in arable land. In addition, debate exists regarding the shift of pastures or rain forest in Brazil, Malaysia and Indonesia to other land uses, where arable has been devoted to sweet corn and palm oil fuel crops.

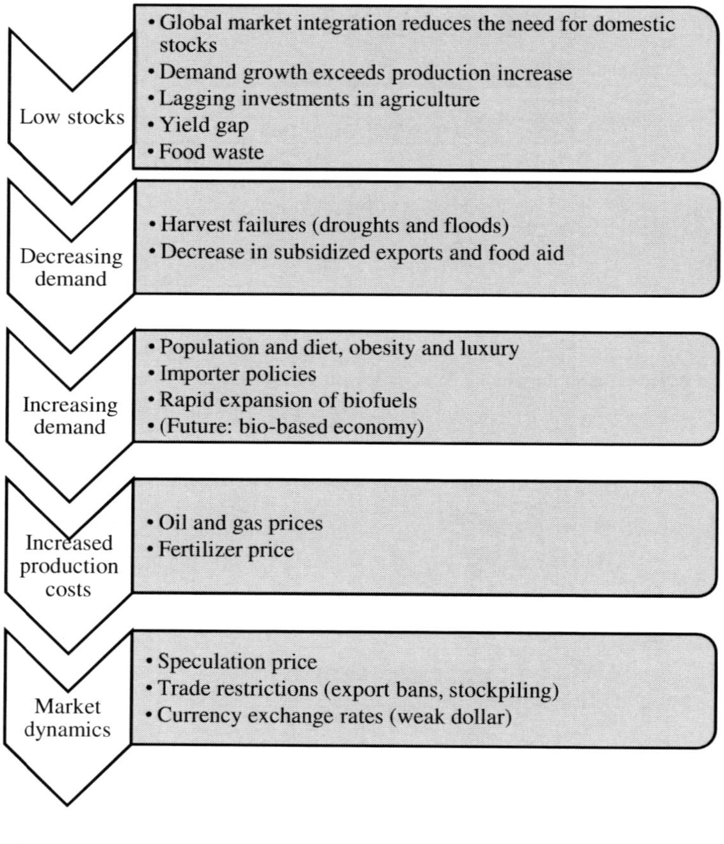

Figure 3.6. Risk factors for food price increases (OECD-FAO, 2009).

World biofuel production has increased by a factor of seven during the last ten years. Uncertainty always exists when weighting the contributions of individual factors. For example, market dynamics are considered as factors that increase food prices. However, discouraging speculation in the food stocks market may increase the risk of price spikes. Environmental risks reached a global scale during the 2007-2008 world food price crises (OECD-FAO, 2009). The government of Brazil claims that the sugarcane based ethanol industry was not instrumental in the 2008 food crisis. Meanwhile, a World Bank policy study released in July 2008 concluded that large increases in biofuel production in the US and EU were the main causes behind steep increases in global food prices. However, this report was corrected by the World Bank, who previously overestimated the contributions of biofuel production, because the effects of biofuels on food prices were not as large as originally predicted. However, the use of commodities by financial investors was potentially responsible for the 2007/08 spike. A 2008 independent study conducted by the OECD also indicated that the impacts of biofuels on food prices were much smaller (Figure 3.6).

3.6. The Impacts of Biofuels on Land Use

Increased biofuel production has been criticized for its potential impacts on food availability. However, it is feared that increasing demands for cropland will result in deforestation and grassland conversion (Biomass Research, 2013, Johannes et al., 2014). Carbon releases from associated land use change are used to undermine reductions in greenhouse gas emissions to the environment. The main criticism is based on the expected impacts of biofuel production that will result from the introduction of dedicated biofuel policies following 2005. The problem of land use change due to biofuel production is discussed in the following sections.

3.6.1. Land Use Change

Another factor of concern is land use change. An example of land cover and land use change is shown in Figure 3.7. China, Brazil, and the USA are referred to as the most important countries regarding wide land areas for agricultural use. This concept uses the MCI (Multiple Cropping Index) as an indicator for calculating the cropping intensity, which is expressed as the sum of all harvested crop areas divided by the total area of arable land. The MCI introduced measuring methods for determining the cropping intensities of tropical farming systems and was calculated for temperature regions (OECD-FAO, 2009). Examples of using the MCI in different study areas resulted in numbers of 0.53 in South Africa, 1.45 in China, 0.8 in Brazil, and intermediate values in the USA, EU, Indonesia, Malaysia, and Mozambique.

The land use for biofuel expansion was calculated by dividing the increased biofuel production presented in Figure 3.7. For example, the Biomass Research Report of 2013 mentioned that biofuel expansion resulted in the use of an additional 25 million ha of cropland in the USA following 2000. Of this land, 11 million ha were allocated to co-products and 14 million ha were used for net biofuel. Over 85% of this area is located in the USA. Since 2000, several countries have indicated decreasing biofuel production areas (by 9 million ha). The total agricultural area in the USA, EU, China and South Africa has decreased by 31 million ha. Moreover, the net global loss of cultivated areas for other purposes reached 48 million ha. Generally, losses in agricultural area have exceeded net biofuel expansion (Figure 3.8).

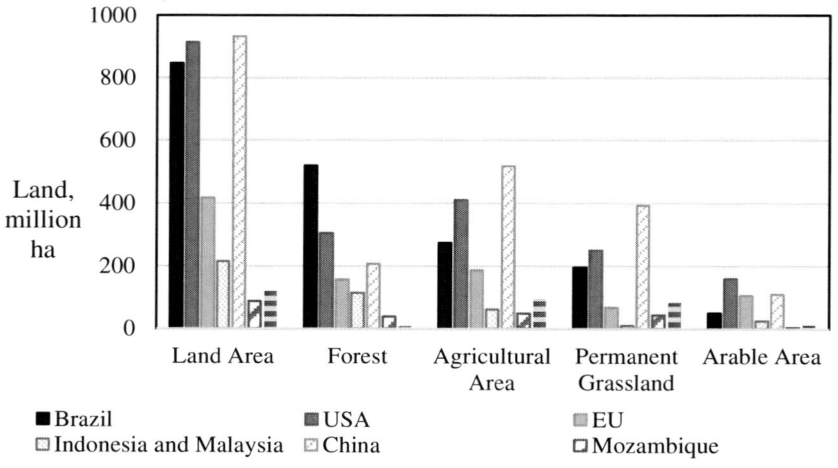

Figure 3.7. Land cover and land use (Biomass Report, 2013).

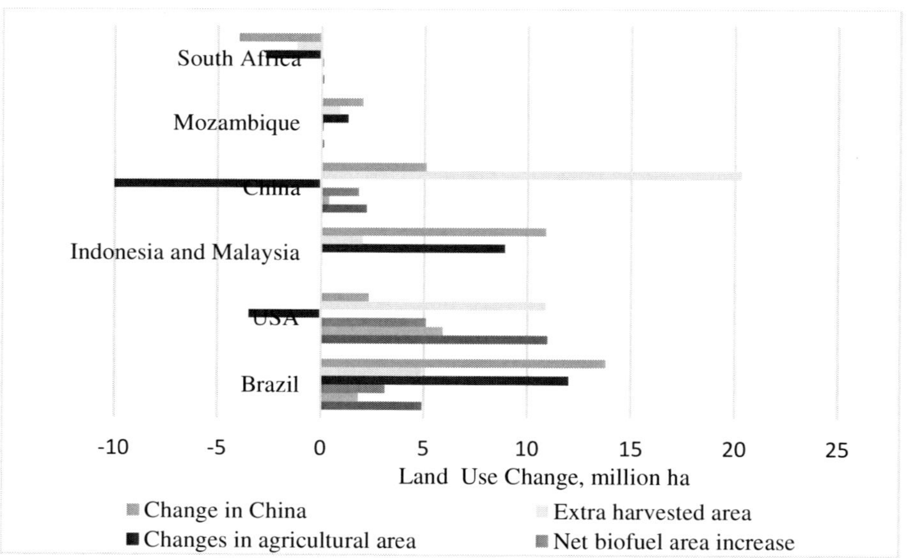

Figure 3.8. Net changes in land availability (Biomass Report, 2013).

3.6.2. Impacts on the Environment

Regarding the use of raw materials for biofuel production, it is difficult to determine the direct and indirect impacts of land use changes. Direct influences have been covered extensively in existing biofuel legislation. How biofuel producers should be responsible for the impacts that change land use when producing raw materials can result in indirect consequences that affect other players.

The total emissions from biodiesel are high relative to the emissions from fossil fuels when considering how each direct and indirect land use changes emissions. Natural gas is worse than the "filthy fuel" coal, which was once considered to result in global warming effects because of methane gas outputs and cooling effects related to aerosols from burning of coal. Pure electrical vehicles are worse than petroleum-burning hybrids when considering their overall lifetime and costs/benefits. In Germany, it has been projected that 10-11 million hectares of agricultural land would be required to provide biomass for biofuels by 2030. Given current trends in land use, the majority of this land would be located outside of Germany, and most feed stocks, such as vegetable oil from land and soy from Brazil, would be imported (UNEP, 2011).

Indirect land-use changes could considerably compromise GHG savings from growing biofuels. In this simulation, the expansion of 121,970 km^2 of rangeland into forestland and 46,000 km^2 of rangeland into other native habitats due to the expansion of biofuel croplands is considered (Lapola et al., 2010). Sugarcane ethanol and soybean biodiesel would be responsible for 41% and 59% of this unplanned deforestation. Planting oil palm instead of sugarcane or soy on pasture lands would result in some direct deforestation (approximately 300 km^2) but would significantly reduce emissions from ancillary land-use changes due to the crop's substantially higher oil yield. Different crops were tested that could serve as feedstocks for fulfilling the demands for biodiesel in Brazil, which indicated that oil palm would result in the fewest land-use changes and the lowest associated carbon debt. Because of its high oil yield, approximately 4,200 km^2 of oil palm cropland would be necessary for fulfilling the demands of Brazil for biodiesel in 2020. In comparison, 108,100 km^2 of cropland would be required for soybean (Lapola et al., 2010).

3.7. Agricultural Expansion and Intensification

Dealing with expansion and intensification of agriculture from an ethical point of view includes considering the effects on the environment. Different agricultural practices used in intensive systems affect the environment in different ways, as briefly noted in the following subsections.

3.7.1. Decreases in Biodiversity

Increases in productivity have resulted in rapid intensification of agricultural practices within the last decades. This increase has resulted in a widespread decline in farmland biodiversity. In fact, different types of species have observed significant decreases in their populations. Birds are among the species that have been affected by agriculture. For example, in the UK, the populations of ten farmland bird species decreased by ten million breeding individuals during the past 20 years (Benton, 2003). Decreasing populations have also affected many other taxa, including mammals, arthropods, and flowering plants. This decline is linked with agricultural practices. For example, in Europe, the change in cereal yield accounts for 30% of the decrease in farmland bird numbers. This result is primarily due to losses of habitats and feeding resources.

3.7.2. Agricultural Practices

Tillage and drainage systems are important for maintaining sustainable soils. Tillage alters many aspects of the soil's physical environment, including the soil water content, aeration, compaction, porosity and temperature (McLaughlin, 1995). These processes render the soil susceptible to wind and water erosion and can lead to de-vegetation, erosion, and desertification. Again, drainage is used to evacuate excess water from wetlands to convert them into agricultural fields. Over half of the original wetlands of southern Canada have been lost, of which approximately 85% of them were lost through agricultural land drainage (Weersink, 2001). Marsh drainage to increase or improve agricultural land has negatively affected true marsh-dwelling fungal species due to direct habitat loss and because of increasing pollutant loads and sediment inputs in drainage water.

3.7.3. Agrochemicals

Generally, two types of agrochemicals are currently used in agriculture, pesticides and fertilizers. Pesticides, which include insecticides, herbicides, fungicides and other chemicals are used to kill certain organisms. Although some degree of selectivity is required for targeting crop enemies, significant declines in the populations of non-target species also result from pesticide use (McLaughlin, 1995). Numerous studies have indicated direct and indirect elimination of birds due to insecticides. Industrially manufactured inorganic fertilizers are popular, easy to transport and spread, relatively odorless, can be used precisely and are effective. In contrast, these types of fertilizers have been linked with greater foliar crop disease, which can influence the vegetation structure and biodiversity of vertebrates and invertebrates and harm human health. In fact, excess nutrients can result in pervasive effects on terrestrial and aquatic biota through eutrophication (Dalton and Brand-Hardy, 2003).

3.8. Trade-off Solutions

In recent years, most global incentive schemes were reformulated as mandates to demand a certain volume of biofuel. With the use of long-term and stable biofuel policies, the potential for markets to predict the proportions of feedstock used for biofuels is required. With the use of long-term and stable biofuel policies, markets must be capable of predicting the maximum proportion of feedstock that will be used for biofuel production. These policies reduce fluctuations in crop demand because portions of the entire demand become predictable. Biofuels are produced from a broad and variable feedstock. The values of feed stocks in every offer chain are limited by competitor choices. For example, if rapeseed production is high, the use of biodiesel from palm oil will becomes more attractive. Thus, this process would indirectly limit a certain volume of biofuel production.

In the future, innovations in alternative technologies must increase to prevent the utilization of biomass reserves and food supplies. Biomass technologies that offer opportunities for development include lignocellulose and algae. Currently, the costs of biofuel production from 2nd and 3rd generation technologies are still considered expensive and are not feasible. First generation biofuels sometimes refer to bioethanol made from sugar-rich (sugar beet, sugar cane, or sweet sorghum) and starch-rich (corn, wheat, or cassava) crops. Again, biodiesel is formed from seed crops (soybeans, sunflower, rapeseed, and palm) or animal materials. In most cases, these feedstocks may be used as food and feed. Non-edible and ligno-cellulosic biomass are major considerations for second-generation biofuels (corn stover, husks, stalk, cane bagasse), forestry residues (thinning, treetops and branches), perennial grasses (switch grass and Miscanthus), short rotation coppice (willow or poplar) and municipal waste. Biofuels that do not compete with food crops or with lands are referred to as third generation biofuels. Typically algae-based biofuels fall within this category (Figure 3.9). The term third generation biofuel has just entered mainstream research and refers to biofuels derived from algae. Previously, algae were considered as second-generation biofuels. However, it soon became apparent that algae can produce much higher yields with lower resource inputs than other feedstocks. Food security must be achieved first for the sustainability of life when considering agriculture. The replacement of edible oil crops for biofuels with non-edible oil crops such as algae and lignocelluloses could reduce risk through technology transformation.

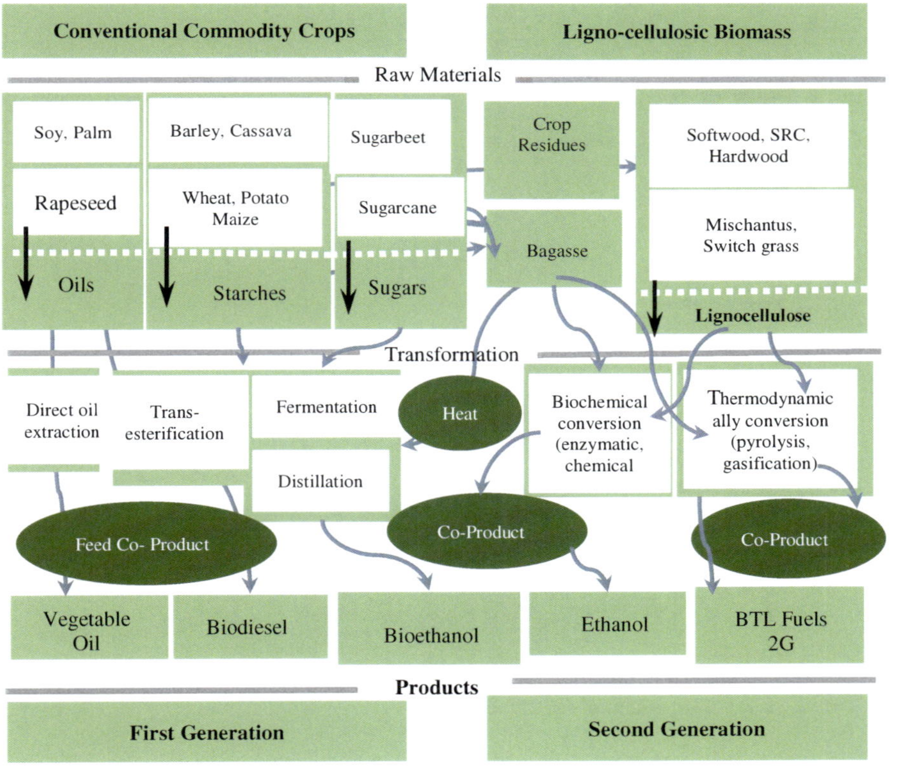

Figure 3.9. Generation of Biofuel Technologies (FAO, 2013b).

The energy potential of biofuels may be high. However, biofuel production results in many different issues globally. Recently, major issues in food price inflation and ecological destruction have been derived from the expansion of biofuel production. The bottom line is used for discussion regarding the fundamental background of biofuel production and to verify the contributions of biofuel production for the next generation.

3.9. Intergenerational Ethics

Intergenerational Ethics (IE) for the next generation are the responsibility of the current generation. Here, we discuss the implications of expanding Biofuel Production (BP) from the viewpoint of environmental recovery. The field of ethics (or moral philosophy) involves systematizing, defending, and recommending concepts of right and wrong behavior. We must deal with ethics as an issue of human to human relationships. The golden rule of ethics is "you should not dislike others". Intergenerational ethics should be considered as one form of ethics based on its golden rule. However, many discussions regarding the formation of intergenerational ethics are necessary because we must discuss the future targets of humanity (not the existence of the other end of ethics).

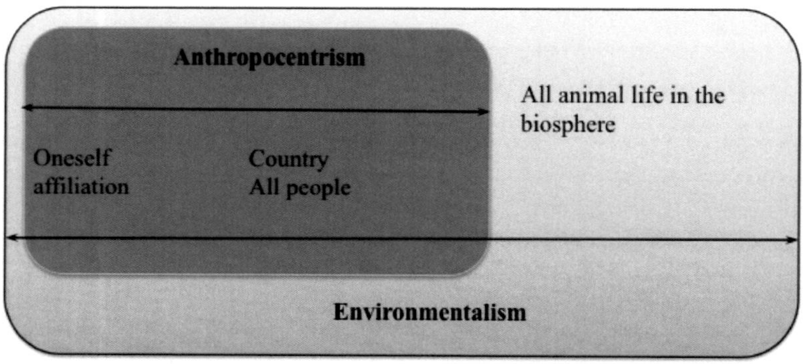

Figure 3.10. Axes of anthropocentrism and environmentalism.

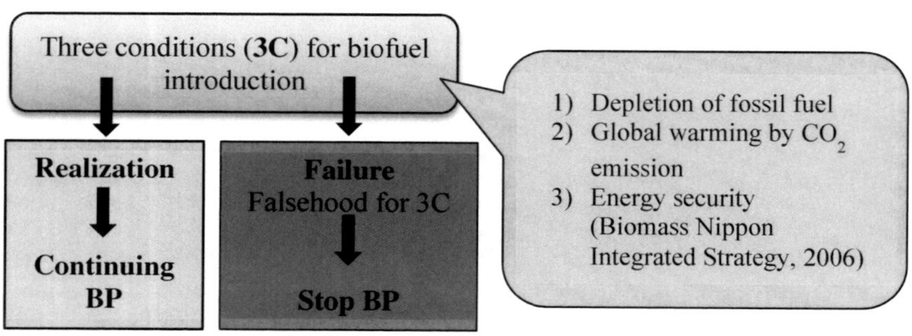

Figure 3.11. Axes for continuing or halting biofuel production.

Major issues of inflated food prices and ecological destruction have resulted from the expansion of biofuel production. These issues influence our generation and will influence future generations for a long time. Thus, it is important to discuss biofuel production relative to agriculture, forestry, natural forests, and the environment. We cannot judge whether biofuel production, agriculture, wood production, or nature preservation are important for advanced and developing countries. Therefore, the rights and wrongs of biofuel production should be discussed from an intergenerational ethics viewpoint using a scientific approach.

Two axes are set for sorting out background contradictions in ethics to discuss intergenerational ethics. One axis is the viewpoint of anthropocentrism or environmentalism (Figure 3.10). Anthropocentrism is a principal of affluent society bequeathing the environment to the next generation. In contrast, environmentalism is the principal of bequeathing affluent natural environments to the next generation. Other axes include the viewpoint of existence for three conditions to promote biofuel production, including depletion of fossil fuels, global warming by CO_2 emissions, and energy security (Figure 3.11). If all conditions do not exist, biofuel production should be stopped.

3.9.1. Forming Conditions of Intergenerational Ethics

Using axes categories of anthropocentrism or environmentalism and continuing or ceasing biofuel production, the forming condition of intergenerational ethics are shown in 4 quadrants (Figure 3.12). The quadrant of anthropocentrism and continuing biofuel production (1) in Figure 3.12 indicates that the destruction of nature for the life of the current generation is their right for making a living and survives due to biofuel production. Thus, discussions of intergenerational ethics are not necessary. The quadrant of environmentalism and continuing biofuel production (2) indicates that the gap between advanced countries and developing countries has widened, which is an example of a land rush (destroying natural forest area to extend biofuel field in developing countries). The quadrant of anthropocentrism and stop biofuel production indicates that environmental recovery is required. If the three conditions of fossil fuel depletion, global warming by CO_2 emissions, and energy security did not existed, biofuel production would be meaningless. This result also occurs for the quadrant of environmentalism and stopping biofuel production (4).

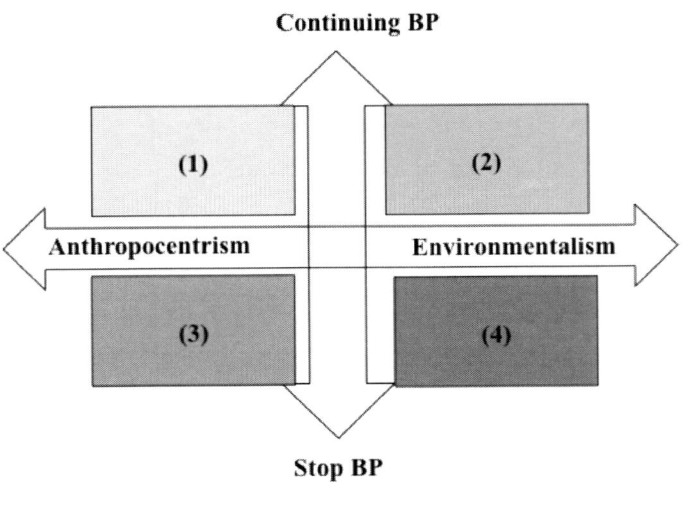

Figure 3.12. Forming the conditions of intergenerational ethics in four quadrants based on two sets of axes, anthropocentrism vs. environmentalism and continuing vs. stopping biofuel production.

3.9.2. Environmental Recovery

The carbon sink potential was studied to clarify the intergenerational ethics for biofuel production from the viewpoint of aboveground and belowground biomass accumulation for 50 years in secondary forests growing on abandoned agricultural lands (Figure 3.13). Data from the literature were used to clarify patterns in aboveground and belowground biomass in the tropics. If the amounts of carbon in secondary forests are as great as those in primary forests in 50 years, no issues should arise regarding the intergenerational ethics of biofuel production from the perspective of anthropocentrism (Figure 3.14). A review of the literature indicated that the amount of carbon in secondary forests did not reach the same amount of carbon in primary forests within 50 years (Figure 3.15). Thus, it is important to consider intergenerational ethics for biofuel production. Several issues exist regarding the intergenerational ethics of biofuel production if a forest is not recovering as a carbon sink. The background of biofuel production is primarily based on politics and the economy. The intergenerational ethics of biofuel production represent a very complicated problem because each country maintains a different viewpoint of future development. The rights and wrongs of biofuel production result in discussions that should be based on science and wisdom to maintain and create a wonderful world for future generations.

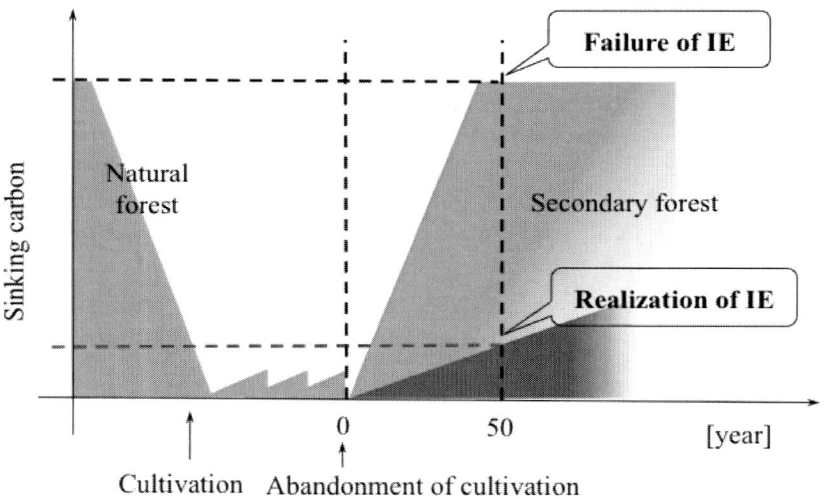

Figure 3.13. Changes in carbon sinks with time.

Agricultural production has significantly changed during the last decades to meet the needs of growing populations worldwide. Intensive cultivation resulting in soil quality degradation, fresh water extraction and the use of agricultural chemicals, (especially dissolved phosphorus) are global issues. Organic agriculture has grown in recent years. In addition, the costs of organic production are favorable for food security. Growing concern regarding chemical contamination with food results in ethical issues regarding health and demands for food security. We must consider food security and meet the need to compromise between the environment, ecology and agriculture. A weak vision of sustainability dominates the methods used for food and energy security rather than a strong vision of sustainability. In the following chapter, we focus on food security and to the indices related to regional vulnerability and food production in the climate change scenario.

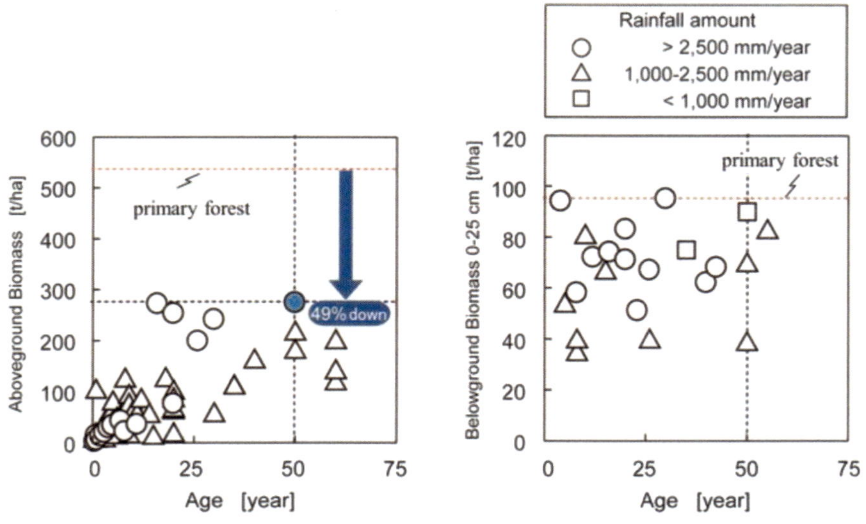

Above ground biomass Below ground biomass

Figure 3.14. Changes in secondary forest biomass after halting biofuel crop production.

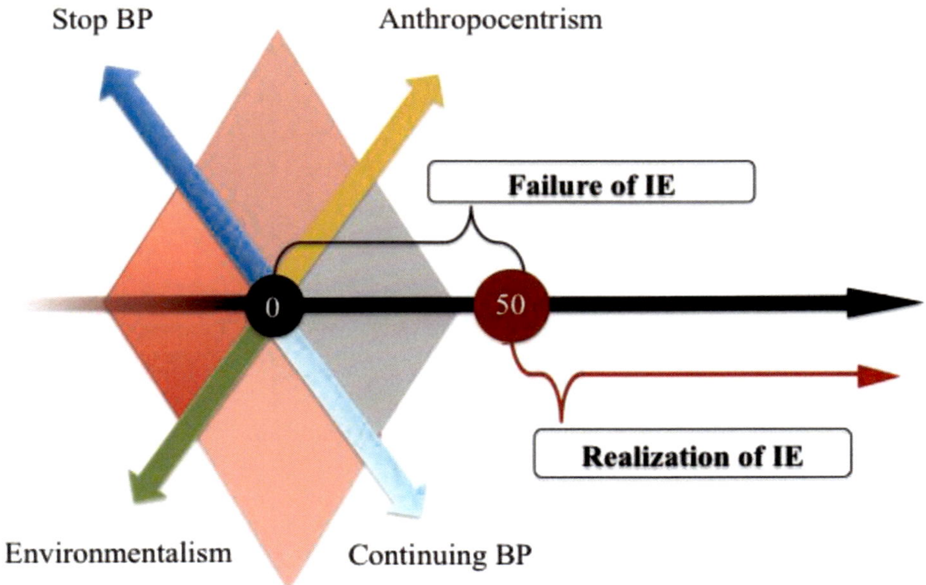

Figure 3.15. Forming conditions of intergenerational ethics on a time series.

Chapter 4

SUSTAINABILITY AND FOOD SECURITY: SUSTAINABILITY AND ITS GLOBAL ENVIRONMENTAL IMPACTS ON FOOD SECURITY

When considering agricultural sustainability, food security issues are becoming more important, and food security is currently a major concern regarding agricultural sustainability. The availability of, affordability of and access to agricultural products are all related to each other. Many challenges in the modern world are related to food supply and climate change. Climate change and global warming are causing changes in weather patterns, with excessive rainfall, flooding, and drought in some parts of the world. The Polar Regions are melting, and accelerated warming and rising water levels have been observed. These phenomena have mainly resulted from human activities and development. In addition, CO_2 emissions are a product of fossil fuel combustion and of burning natural gas. A steady supply of rice, wheat, fruit and vegetables is necessary for a growing community. Growing demands for food require that food products are sourced from farther away or from communities that are obliged to move elsewhere. Considering the demands of agricultural, seeds, fertilizer and water are among the world's most important resources. Only 2.5% of the world's water is freshwater, and more than two-thirds of this freshwater is unavailable for human consumption. When the human population increases, water availability per person decreases. The human population has increased nearly tenfold during the last three centuries and has grown by a factor of four during the last century. Agricultural sustainability faces challenges amid threats of lacking resources, climate change and a growing population. Half of the world's population resides in cities. Rural populations are decreasing rapidly as poverty and lack of work push people into urban centers, which require sustainable urban communities. By 2050, approximately three-quarters of the world's population are expected to reside in cities. To meet this pressure, energy demands will remain a challenge throughout the world. Access to energy is a measure of the progress of urbanization and its sustainability. Renewable energy could be supplemented and enhanced by using green energy. A driving force is needed that could change attitudes towards self-sufficient societies and eco-environments. Therefore, sustainability and food systems play a significant role in enhancing quality of life without compromising the vision of sustainability. Sustainability from agricultural, environmental, and ecological perspectives is a primary concern for agricultural practices. Agricultural production and soil and water use in the context of climate change have received attention throughout the world. A brief discussion regarding this focus is presented. This chapter addresses food systems and food security issues in climate change scenarios. Many models have been introduced to understand the advanced interconnections that exist within the food system.

4.1. The Food System

The food system includes food production, processing, distribution, marketing, acquisition, and consumption. In addition, a healthy, properly nourishing system includes environmental quality, economic vitality, human health, and social equity. The idea of food systems is often used in agriculture, food science, nutrition, and pharmaceuticals to explain the complicated set of activities required for providing food for sustenance and with nutrients for maintaining health. The series of transformations within the food system is commonly characterized using the words "from field to table" or "from land to mouth" (Kneen, 1989). Each aspect of the food system is ultimately related to access and the availability and utilization of nutritious foods. The food system can be distilled down to community food systems, which include everything from farm to table. A community food system is one in which food processing, allotment and distribution are integrated to enhance the environmental, economic, social, and nutritional health of a particular place. Community food systems are also referred to as "local" or "regional" food systems (Rutgers, 2014) and are considered customized approaches for answering food questions and valuating economic, environmental, and social sustainability based on community efforts. In the holistic view, the concept of community encompasses the global context in which relationships are built around food.

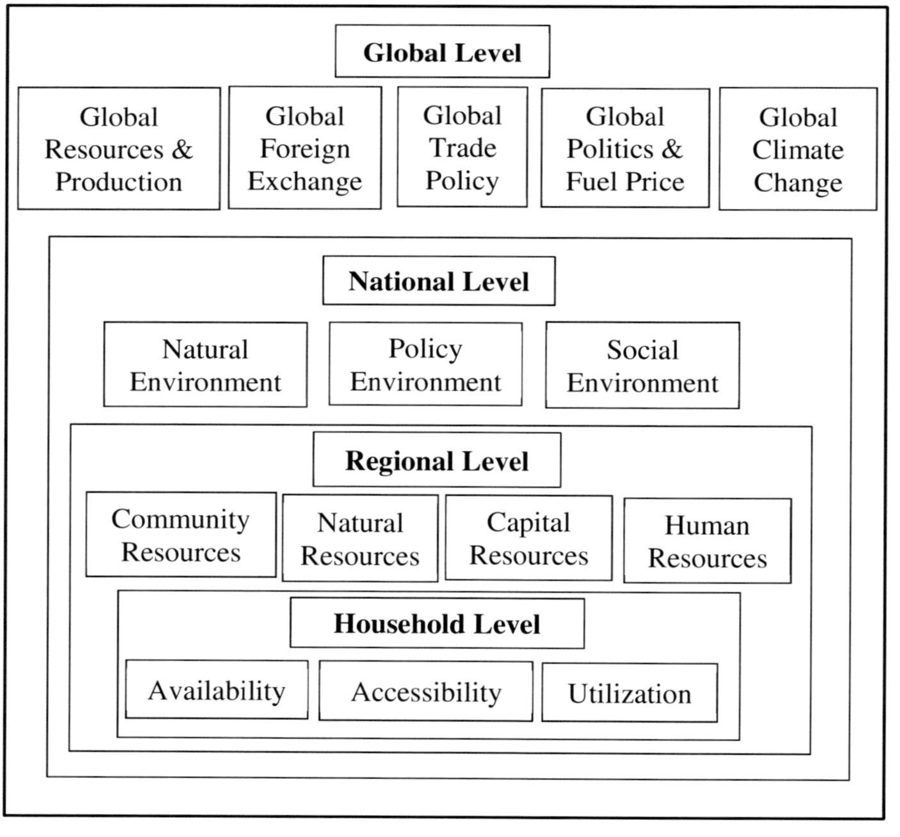

Figure 4.1. Food System Model.

The food system defines the boundaries of the food network that contain the food chain, which is represented using models based on flow diagrams at the global, national, regional, and household levels (Figure 4.1). The world food system offers an important summary of the relationships of food at the national and international levels (Fischer et al., 1996). This system identifies potential gaps and excess within food systems, their causes, and methods of addressing them through more robust resource use. In addition, the world food system attempts to identify any actual or potential environmental impacts caused by food production. Food systems are diverse, complex, and dynamic, and each level of a system is related to the other levels. Control of the world food system should be globalized and industrialized. However, a more community-oriented approach should be used to develop more local, personalized and sustainable food systems. At the national level, policy, social and environmental factors affect food system dynamics, which are driven by regional factors such as community, capital, natural, and human resources. At the household level, the food system is reflected by food security indicators, such as availability, access, and utilization (Riely et al., 1999).

4.2. Food Security

Food security means that all individuals have physical and economic access to adequate amounts of nutritious, safe, and culturally appropriate foods that are produced using environmentally sustainable and socially acceptable methods (Food Security Network, 2014). Food security is achieved when adequate food (quantity, quality, safety, sociocultural acceptability) is accessible to and satisfactorily used by all humans, and governance and political stability influence food security. The most widely accepted current concept of "food security" was given at the 1996 World Food Summit: "Food security is guaranteed when, at all times, all people benefit from economic, social and physical access to sufficient, safe and nutritious food to meet their nutrition needs and food preferences, so that they can lead an active and healthy life." This definition fully encompasses the dimensions of food security based on four characteristics, food availability, food access, the stability of food availability, and nutrition. The phrase "all people, at all times" is integral for defining food security, and is essential for achieving national food security objectives. Society still faces a number of food security issues, specifically, the continuing shift towards food insecurity. Social factors are currently recognized as one of the causes of food insecurity in many parts of the world. Additionally, natural hazards and stresses, such as droughts, market failures, political instability, and institutional weakness have resulted in food security concerns (Devereux, 2000, Corral et al., 2000, Devereux and Maxwell, 2001).

4.2.1 The Dimensions of Food Security

The dimensions of food security and their variants influence social, political and natural environments (Figure 4.2). Food availability depends on domestic food stocks and commercial fair trade between markets and consumers.

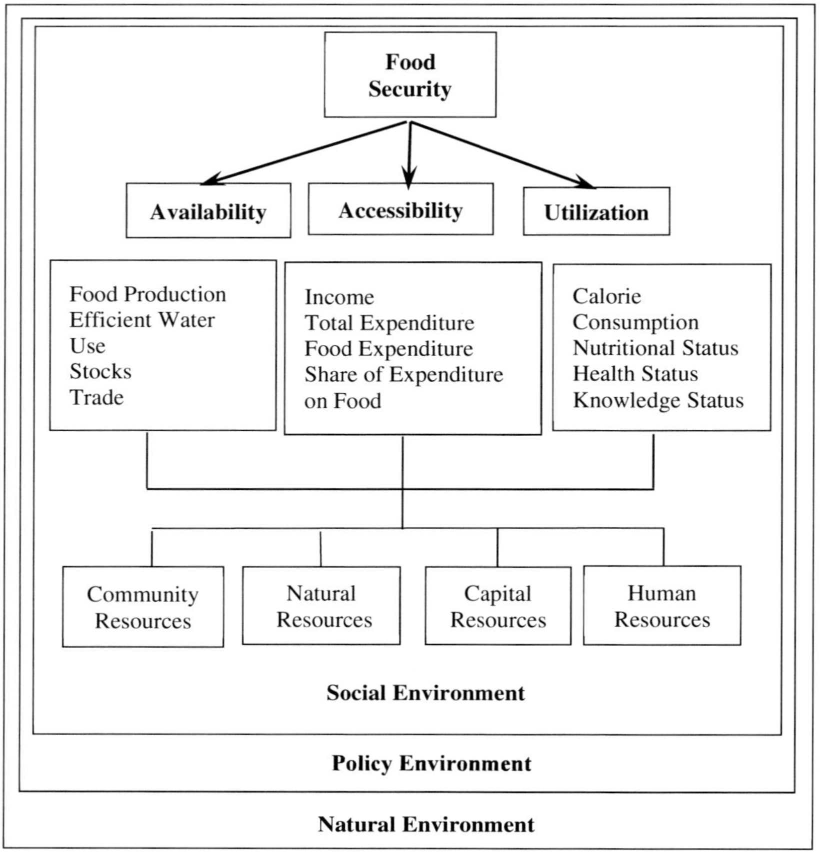

Figure 4.2. Dimensions of food security.

In this regard, food security monitoring can help to identify and understand the effects of food insecurity and hunger on societal well-being. Thus, global, regional, national, and local food security is an essential, universal dimension of household and personal well-being. The World Food Summit of 1996 defined food security as "when all people at all times are given access to sufficient, safe, nutritious food to maintain a healthy and active life" (WHO, 2014). Inadequate food security and nutrition results in economic imbalances and negatively affects people's livelihoods and economic capabilities.

4.2.2 Food Insecurity

The concept of food insecurity is complicated and multidimensional and follows successive stages as it becomes more severe. Each stage consists of characteristic conditions and insufficient food experiences to meet the essential needs of families and communities. A population is considered as acutely food insecure when a large reduction in its primary food supply occurs or when the prevalence of malnutrition is abnormally high for the time of year. Food insecurity exists if individuals do not have adequate physical, social or economic access to food.

4.2.3 Influences on Food Insecurity

According to the World Bank, approximately 44 million people were forced to live below the poverty line in 2008 because of spikes in global food prices. The most significant aspect of this crisis is that it mainly occurred in poor countries. In 2014, the world population increased to 7 billion, and the average population growth rate was 1.2% in 2013. Although the world's current food production is considered as sufficient for feeding all of its inhabitants, food security could become vulnerable as the population increases. According to FAO data, the current level of worldwide food production provides 17% more calories per person than 30 years ago. However, within the last 30 years, the population increased by 70%. One difficult task is overcoming the physical, social, political, and economic barriers to food access for the people who need it most. Food security indicators have been divided into three categories: affordability, availability and food safety, and quality. The descriptions of Global Food Security for these categories are discussed in the following section.

Affordability is measured by the following six indicators: food consumption as a share of household expenditures, the proportion of the population under the global poverty line, the gross domestic product per capita, agricultural import tariffs, the presence of food safety net programs and access to financing for farmers.

Availability is a composite indicator of sufficient food supply, public expenditures on agricultural research and development, agricultural infrastructure, volatility of agricultural products, political instability risk, corruption, urban absorption capacity and food loss. The third factor, food safety and quality, entails indicators such as diet diversification, nutritional standards, micronutrient availability, protein quality, and food safety. Sub-indices for the availability, safety and quality of certain foods are also used. By combining all of the indicators, the results are used to analyze current world food security. The global food security index now considers 28 individual indicators of the three core perspectives of food security for 109 countries, including developed and developing countries. Here, a group of experts has worked together to address all of the factors that influence how countries ensure food security for their populations. These experts attempted to compile all of the related aspects and made the resulting data available to assist policy makers, students, and researchers in analyzing and recommending policies. This work allows countries to compare themselves with countries that are economically similar with differences in food security. The data permit each country to focus on the areas in which they are doing well and in which the most improvement is needed. In addition, countries are also ranked according to an analyses of their individual food insecurity risks. These data are also useful for individual stakeholders of countries for addressing different shocks and taking advanced measures to improve food security. The main purpose of the global food index is to identify the problems that cause food insecurity, of which one important driver is food price. Thus, the global food security index incorporates a food price adjustment factor.

4.3. Household Food Insecurity

The nutritional status of each member of the family depends on the availability of food in the market, especially food of sufficient quality and safety, and each person should have good health. The United Nations acknowledges that healthy, well-fed individuals result from successful social and economic development and represent an important factor in the development of a society. Poor individuals spend larger proportions of their income on food, and a small numbers of farmers derive most of their income from producing food in developed regions. This implies that changes in food costs could have large effects on the welfare of farmers. For instance, if a farmer receives his expected price, he and his family can survive and he may have sufficient surplus to feed others. However, when food costs are high, poorer consumers suffer. At the family level, the primary indicator of food security is a households' access to and ability to provide food based on necessity. A second keystone indicator is increased biofuel production, which could threaten food security and value as well as family incomes; thus, larger families may be able to purchase additional food.

4.4. Measuring Food Insecurity

In recent years, the awareness that food insecurity should be analyzed in an exceedingly dynamic context has increased. It is important to consider current inadequate nutritional outcomes, specifically for identifying individuals, households, and communities that are most at risk of suffering in the future. In addition, no established consensus is available in the literature regarding the most appropriate approach for analyzing this vulnerability. Furthermore, most analyses of vulnerability have focused on poverty rather than on food insecurity. Traditional approaches tend to emphasize the roles of assets in reducing vulnerability. Even more significantly, some of the most common methodologies used to analyze vulnerability are static in nature and fall short of appropriately assessing the dynamic nature of vulnerability. The influential entitlement approach of Sen (1988) links vulnerability with inadequate access to assets, including intangible assets such as social capital. However, access to assets does not guarantee that the assets will be utilized in an effective way to reduce vulnerability. The UK Department for International Development (DFID), for example, developed its vulnerability assessments in terms of household assets and the activities required to maintain or sustain livelihoods (DFID, 2003).

Vulnerability is a forward-looking concept based on the notion that the food security outcomes of households are a realization of a stochastic process that is conditional for a number of risk factors at different levels. Vulnerability can be altered by using endogenous strategies implemented by households, communities or public institutions. In contrast, the World Bank uses a risk-based approach for assessing household vulnerability (Holzmann, 2000). The bank's "Social Risk Management" framework considers the sources of vulnerability and the communities' ability to manage the associated risk. The emphasis is largely on minimizing risk exposure, although one major weakness in the approach is a failure to consider the risks that stem from insufficient ownership of or access to assets.

Food insecurity should be defined according to its different dimensions, including availability, access, stability and utilization (Pingali et al., 2005). Hence, an appropriate conceptualization of vulnerability should consider all of these aspects. To develop an index, all of the above mentioned indicators could be combined with suitable anthropometric measures, and the index could be constructed using classical statistics techniques, such as principal component or factor analysis (Morrison, 1967).

If we let y denote the food security indicator that summarizes a household's food security according to different dimensions. Then, a households' vulnerability to food insecurity can be defined as the expected welfare loss that is associated with an inadequate food security indicator value and is conditional on a number of household characteristics, such as the risk management policies of public institutions and factors beyond the control of households and public institutions (Scaramozzino, 2006). These variables are denoted by the vector x. Thus, a household vulnerability indicator could be expressed as follows:

$$V_1(y^*,x) = \int_0^{y^*} W(y) dF(\frac{y}{x}) \quad (4\text{-}1)$$

where y^* is a critical benchmark value for the food security indicator, x is a vector of the current household and community-specific conditioning variables, $W(y)$ is the household welfare function, and $F(y|x)$ is the conditional distribution function of the future food security indicator, given the vector of conditioning variable, x. The vector x could include the current food security indicator value. Hence, this definition allows the possibility that a household, which is not currently suffering from inadequate nutritional outcomes might do so in the future and that a currently food-insecure household might overcome its vulnerability (Scaramozzino, 2006). Thus, this definition captures the dynamic, forward-looking nature of food insecurity as well as its stochastic properties.

The above general definition $V_1(y^*,x)$ is based on an assessment of the welfare losses associated with food insecurity. As a special case of the above definition of vulnerability, it is possible to consider an alternative indicator that measures the conditional probability of inadequate nutritional outcomes. This alternative indicator can be expressed as follows:

$$V_2(y^*,x) = \int_0^{y^*} dF\left(\frac{y}{x}\right) = F(\frac{y^*}{x}) \quad (4\text{-}2)$$

Here, the same notation is used as described for the previous indicator. Briefly, the second indicator $V_2(y^*,x)$ measures the probability mass associated with an inadequate food security indicator value. This indicator is consistent with the notion of absolute or relative value-at-risk and could be interpreted as a counter-indicator of vulnerability to food insecurity relative to an appropriate threshold or benchmark y^*.

4.5. Food Security and Strategies

In Figure 4.3, the FAO approximation of undernourishment around the world is shown according to developed and developing regions. Furthermore, the data is also presented for Asia, Africa, Latin America, and Oceania. According to this figure, 98% of the world's hungry live in developing countries, including 63% in Asia and 25% in sub-Saharan Africa. These data are alarming because this crisis occurred despite global concern and actions for mitigating hunger and improving food production. Thus, it is enlightening to observe that many issues can cause food insecurity, but it is apparent that physical factors have always been the main cause of serious hunger. For example, human activity can affect food prices and people's general ability to purchase food even when ample food is available in the market. Another important aspect is the free market, which was introduced by the EU, the USA and certain institutions for promoting competition and new open global food supply markets in which developing countries can participate. However, this effort has received a number of criticisms, including that free markets directly affect European and American subsidized food supplies and less expensive imported foods create challenges for local food production. It is difficult for local farmers to compete with food prices, which results in lower farmer income and leads to food insecurity. Another factor is the fluctuation of food prices in the global market, which change with global conditions, such as oil prices and the trade of goods and services. Food prices increase when food is traded as a commodity, which also decreases local food supplies and results in food insecurity. Generally, food budgets of families in developing countries are always greater than those of families in developed countries. Therefore, increasing food prices directly affect food security in developing countries and represent the primary cause of an increase of approximately 14% in the number of hungry people between 1995 and 2009. Increasing food prices have also resulted in undernourishment worldwide (Figure 4.3).

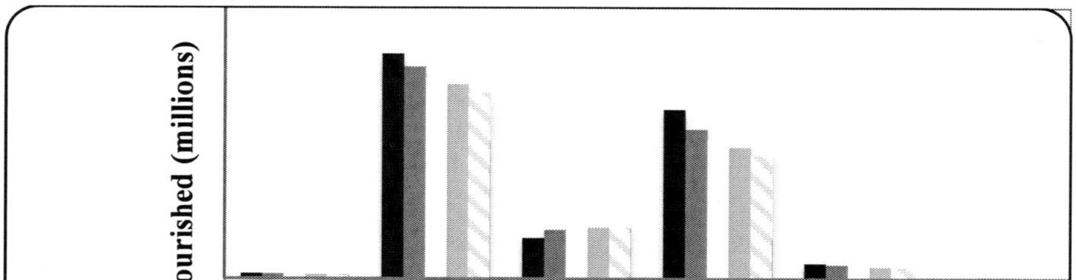

	Developed Regions	Developing Regions	Africa	Asia	Latin America and the Caribbean	Ocenia
■ 1990-92	19.8	995.5	177.6	751.3	65.7	0.8
■ 2000-2002	18.4	938.9	214.3	662.3	61	1.2
2005-2007	13.6	892.9	217.6	619.6	54.6	1.1
■ 2009-2010	15.2	863	226	585.5	50.3	1.1
2011-2013	15.7	826.6	226.4	552	47	1.2

Figure 4.3. Undernourishment around the world, 1990-92 to 2011–2013, (FAO, 2013c).

Food strategies must consider all of the policies that affect the threefold challenge of shifting production to where it is most needed. Food security requires effective conservation programs that can be and typically are overridden and undermined by inappropriate agricultural, economic, and trade policies. To ensure the long term food security of a nation, more comprehensive approaches and policies must be implemented. The most recent estimates by the FAO indicate that 842 million people (12% of the global population) were unable to meet their dietary energy requirements between 2011 and 2013 (FAO, 2013c). Thus, roughly one in eight people in the world is likely to have suffered from chronic hunger and insufficient food for leading an active and healthy life. In developing regions, where the prevalence of malnourishment was estimated at 14.3% for 2011–2013, approximately 827 million people were chronically hungry (FAO, 2013c).

4.6. THE FOOD SECURITY INDEX

Food security has received more attention in the international forum as a social stability crisis, or in the context of the political and economic stability of states with oversized proportions of low-income inhabitants who are disproportionately affected by food prices. The World Bank estimated that approximately one hundred million people are facing greater poverty because of rising food costs. Approximately thirty countries have faced challenges in political and economic stability. Consequently, social riots have occurred and have resulted in food insecurity. The long-lasting consequences of the recent food crises for food security and nutrition in developing countries have become a topic of wide discussion. Some modest progress has occurred in the form of the international community's commitment to MDGs, which recommend methods for managing hunger and deficiency diseases throughout the world (IFPRI, 2012). However, effective food nutrition systems require attention to specific contexts and specific interventions that can best address the problem within a particular context. In this regard, policy analysis was conducted over a decade to develop and rank food insecurity regions following basic ideas of national food resilience, food production, foreign currency reserves and household food access based on purchasing capacity (Menrolla, 1989). A country is food insecure when its economic system is unable to furnish sufficient food to feed its population because of inadequate economic access to available food (Downing, 1991). Menrolla proposed the Global Food Security Index (FSI), which consists of the following three main components: the National Food Resilience Index (NRI), Household Access Index (HAI) and Performance Index (PI). In the following sections, the three main components of the FSI are discussed briefly.

4.6.1. National Food Self Resilience Index (NRI)

National food self-resilience is outlined as a country's capacity to produce food and a country's capacity to import food that it cannot produce (Menrolla, 1989). The NRI is one of the three fundamental indices that form the basis of this analysis. In accordance with this wider definition, the NRI is composed of per capita food production and gross foreign exchange earnings. Different critics might contend that the means of national food self-resilience should be extended to incorporate variability in domestic food production and foreign exchange earnings similar to the average levels of those indicators. Two monetary indicators are determined. First, the FOODPROD is determined, which is the average annual per capita food production over a two-year period. Second, the FOREX is determined, which is the average annual per capita gross foreign exchange over two years. The other minor indicator is AFOREX, which represents the average annual per capita gross exchange earnings for a particular year adjusted for debt service. Gross foreign exchange comprises the earnings from merchandise exports and services. Generally, NRI can be expressed as follows:

$$NRI = FOODPROD + FOREX \quad (4\text{-}3)$$

4.6.2. Household Food Access Index (HAI)

Economic access to food is theoretically expressed using incomes, such as those from specific livelihoods, and food costs. In addition, the HAI is important because it justifies the premise of this discussion. Because it is difficult to obtain reliable and comparable data for food prices and employment, the HAI is operationally outlined in terms of average gains, food consumption levels and child mortality rates, of which child mortality rates may serve as a proxy for the distribution of economic gain and food consumption (Menrolla, 1989). As such, certain indicators are associated with the HAI. For example, the GNP, per capita GNP, and per capita values could moderately reflect average access to food based on demand. However, these values give no indication of the distribution of financial gains. The CALORIES parameter represents the typical daily calorie consumption per capita, and the INFANTS parameter represents the death or mortality rate per thousand for children under the age of 5 years. The household access index can be expressed as follows:

$$HAI = GNP + CALORIES + INFANTS \quad (4\text{-}4)$$

A country can improve its food security over time, by "performing" well with regard to the indicators, which is evaluated as the performance index of the country. Ideally, the performance index should measure growth or improvement in the five variables that comprise the NRI and HAI.

4.6.3. Performance Index (PI)

A country can improve its food security over time by referencing a few visible indicators. The PI is one of three elemental indices that represent the premise of this analysis (Menrolla, 1989). The PI specifically represents growth in three variables, per capita domestic food production, foreign exchange earnings and GDP. This value is used within the context of growth, whereas GNP was employed in the household food access index. Growth information for average calorie intake, child mortality, and other applicable access variables would enable an analysis of a country's performance for expanding economic access to food. The three variables that comprise the PI include the following: FOODINDX, the index of average annual per capita domestic food production over three years compared with the average production over different three-year periods considering five-year interval population estimates; FOREXIND, the index of average annual per capita foreign exchange earnings over three years compared with the average earnings for the three-year period; and GDPGRWTH, the average annual growth rate of per capita GDP over a five-year period. The performance index can be presented as follows:

$$PI = FOODINDX + FOREXIND + GDPGRWTH \quad (4\text{-}5)$$

The food security index can be measured using five key factors and is used for the two indices. There are a number of ways to relate the five factors with the two indices. The most methodical and impartial approach is to weight each index element equally. However, each variable could also be given equal weight. Using this approach and by totaling a country's overall food security, household economic access to food was found to be 50% more important than national food self-resilience. Thus, the composite FSI can be obtained by combining the two NRI variables with the three HAI variables or weighting each index component equally. Therefore, the FSI can be expressed as follows:

$$FSI = GNP + CALORIES + INFANTS + FOODPROD + FOREX \quad (4\text{-}6)$$

Next, the resultant index can be used for comprehending and evaluating barriers to food security in a variety of ways. A ranking function can be used to rank countries that are currently at most risk, and the ranking order would show which countries need the most attention. Related data on a particular country or region can be extracted from data banks and used to compare economically similar countries. However, in this modern era, food prices are very important aspects of food security. Every day, global food prices fluctuate for a number of social, economic, political and/or environmental reasons. Thus, researchers highly recommend adding a food price adjustment factor to the food security index.

4.7. Food Systems and Price–Environment Interactions

Recent reports have focused additional attention on the links between global environmental change and food security, particularly between global climate change and agricultural systems. The World Bank and other international organizations have focused on greenhouse gas emissions and on the impacts of global climate change and adaptation (World Development Report; 2010, UN Human Development Report; UNDP, 2007). The Food and Agriculture Organization of the United Nations produced a large number of reports on environmental problems and on the state of food and agriculture in the context of environmental services (FAO, 2007, 2008). However, these reports focus more on carbon sequestration and mitigation (Müller-Lindenlauf, 2009) and their impacts on fisheries and food safety. The impacts of climate change on crop yields and food security indices influence domestic and international food prices. The major climate components and food security indices greatly impact the supply and demand of food. In addition, the performance index is a component of this system (Figure 4.4). Global environmental change, land degradation, loss of biodiversity, changes in hydrology, and changes in climate patterns result from greater anthropogenic emissions of greenhouse gases and could have serious impacts on food security, particularly for more vulnerable groups. The speed, scale and consequences of human-induced environmental modifications are on the far side of human expertise. Therefore, policies are required that support food systems (Carpenter et al., 2009; Steffen et al., 2003). In the following section, global environmental change and food security are discussed in more detail in the context of meeting the challenge of climate change.

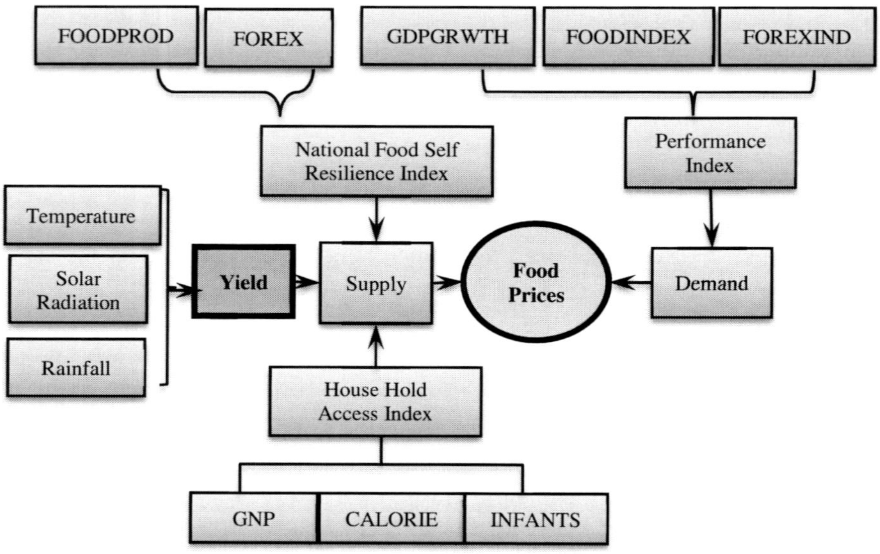

Figure 4.4. The impacts of climate on the food-price system and food security index.

4.8. Global Environmental Change (GEC) and Food Security

The term "climate change" has been utilized in recent years to specifically refer to climate change resulting from human activity and is often used interchangeably with the term "global warming" (EPA, 2009). In addition, GEC is a threat to food security, particularly for people who are engaged in small-scale agriculture for their livelihoods. Simultaneously, agriculture and conjointly connected activities contribute to GEC, for example, by intensifying gas emissions and altering land surfaces. Food security is negatively affected by GEC, although measures to extend food security could increase GEC. This complicated and dynamic relationship is influenced by additional factors. Separate from GEC, food systems are heavily influenced by socioeconomic conditions, such as macro-level economic policies, political conflicts, and other vital drivers (Ingram et al., 2010).

It has always been challenging to ensure food security for the populations in many countries in Africa, Asia, and Latin America, where many people live below the poverty line. Multiple social, economic, environmental, and political issues create barriers to achieving population-wide food security. Moreover, global environmental changes are reflected in climate change, degradation of biodiversity, and loss of agricultural lands, all of which are placing pressure on food security. It is important to understand the impacts of global environmental change on food security. Often, the only concern is agricultural food production. However, climate change can affect food security in many ways apart from crop yields. For example, adaptation is very important in crop production. Many consider that adaptation is only related to the place of production; however, it occurs at many points in a food system. Food production and consumer food supply are elements of food systems, and the availability of food from production is only one dimension of food security (Ericksen, 2008).

Food security is a concern in Africa, Asia and Latin America due to social, economic and environmental stressors. Future global environmental changes, such as exaggerated climatic variability, land degradation, and loss of diversity, could bring unprecedented stresses to food systems and security. To understand the effects of world environmental changes on food security, it is important to target food systems, not simply agricultural production, and determine how changes in temperature, rainfall and solar radiation affect food security. In addition, numerous studies have examined the effects of temperature changes on crop yields. Model simulations of maize production in Africa and other geographic regions expect an overall reduction of 100% by 2055 (Jones and Architect, 2003). In addition, food stability could also be influenced by the increasing frequency and severity of such extreme events, such as cyclones, floods, hailstorms, and droughts (Intergovernmental Panel on Temperature Change, IPCC, 2007). However, food security depends not only on food supply but also on the ability of social units to collect and store food. Increasing temperatures could increase the risks of food spoilage and food-borne diseases. Researchers have found that reportable cases of common food-borne illnesses, such as food poisoning, increase nearly linearly for every degree of increase in weekly temperature (D'Souza et al., 2004; Fleury et al., 2006; Kovats et al., 2004).

4.9. Effects of Climate Change on Productivity

Climate change scenarios have been used to estimate the effects of these changes on crop yields and food trade. A climate change scenario is described as a physically consistent set of changes in meteorological variables based on generally accepted projections of CO_2 (and other trace gases) levels (Fischer G. 2009). The temperature changes in these GCM scenarios (4.0–5.2°C) are at or near the upper end of the range (1.5–4.5°C) that is projected to double CO_2-related warming (IPCC, 1990a, 1992). Crop yields are heavily influenced by weather and climatic parameters. Solar radiation, rainfall, and temperature are actively involved in the biological plant functions known as photosynthesis and respiration (Figure 4.5). Each of these parameters influence yield and vary from year to year. Farmers have now acclimated to adapting their processes to inconsistent and unpredictable weather. However, a high degree of adaptation is still required that specifically addresses local climates, local production and individual knowledge and experience. Although climate change significantly affects agriculture and food systems, several possibilities are available for addressing and improving the situation. The current state of affairs demands an analysis of the impacts of climate change during the coming two decades. This analysis may be inevitable because climate is not the only influencing factor. Other important aspects related to human activity and social, economic and political issues all greatly influence greenhouse gas emissions. Nevertheless, the current situation is already at a point at which some degree of ongoing warming would occur for decades even with greenhouse gas emissions immediate decrease. In addition, sea levels would continue to rise for centuries because the climate would require time to respond to these changes (Gornall Jemma, et al., 2010). Very little information is available in the literature regarding the impacts of temperature changes over these periods. Therefore, the

IPCC presented climate projections for 2020 and 2050 with the corresponding impacts on crop, forest, and livestock productivity throughout the world (Figure 4.6).

Because climate change expatiates the environmental and socioeconomic drivers of food insecurity, a clear global picture of where the threats are greatest is needed, and definite, practical interventions should be implemented to increase resilience in these areas (Beddington et al., 2009). For example, some of the world's major food-producing regions lie in mega-deltas that are threatened by high rates of saltwater intrusion. In Africa, broad gaps exist between the actual and potential crop yields. In Europe, North America and Asia, intensive agriculture, food storage limitations, retail systems and eating habits produce high per capita greenhouse gas emissions and food waste.

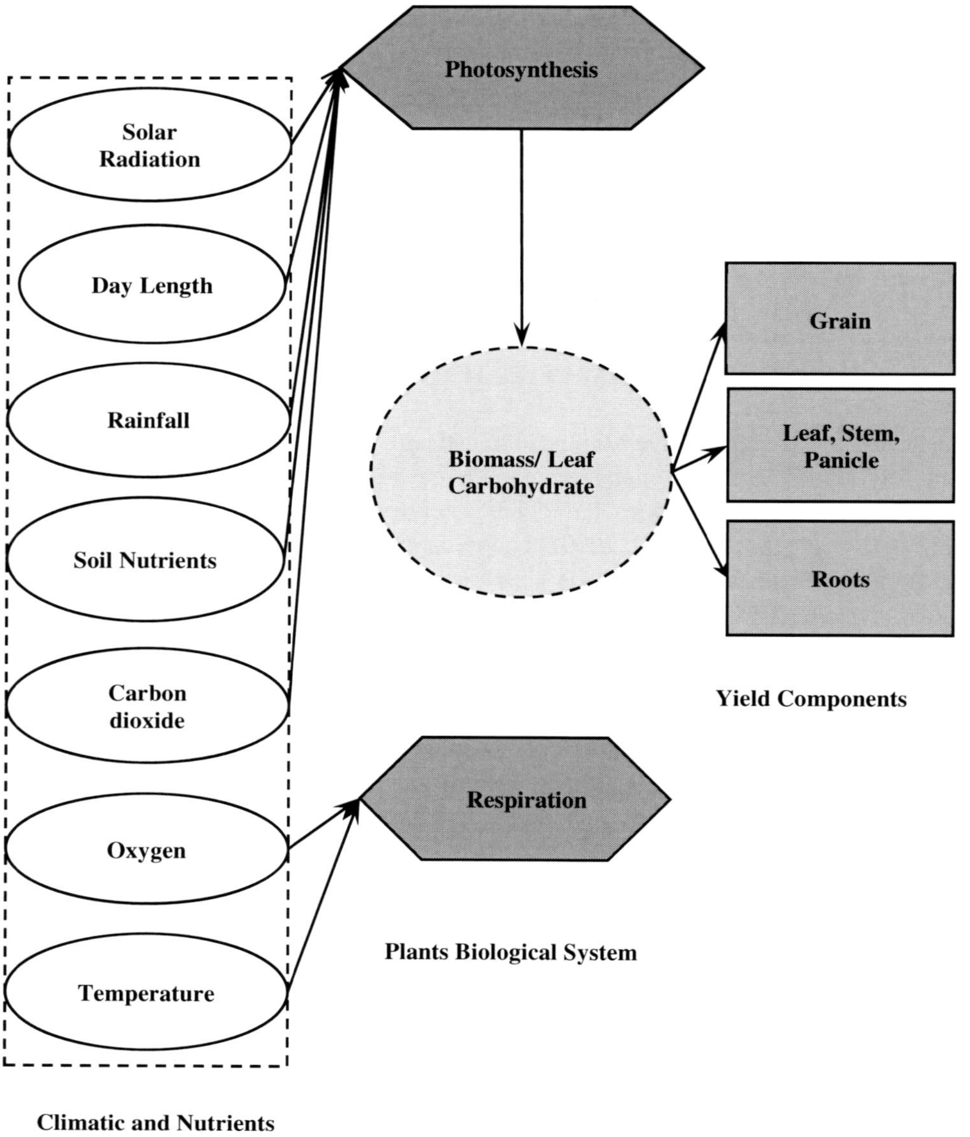

Figure 4.5. Roles of climatic parameters on plant yields.

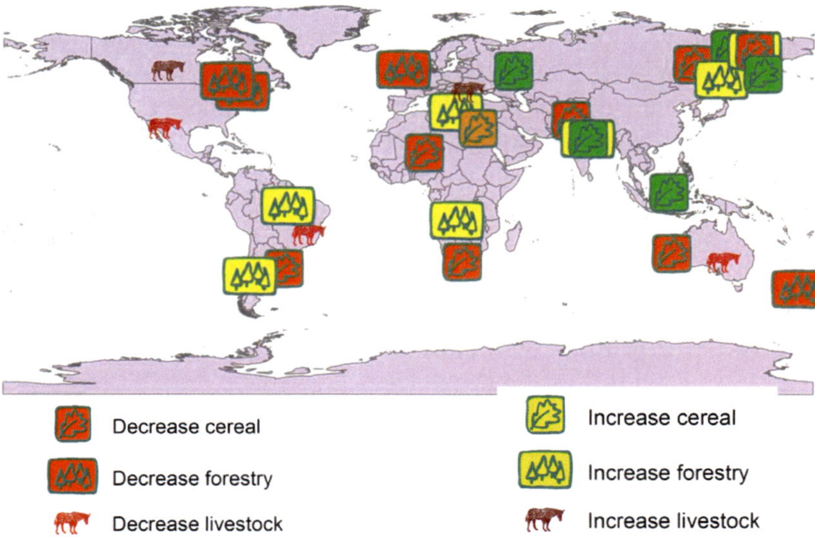

Figure 4.6. Impacts of climate change on productivity (IPCC 2007).

4.10. Challenges to Food Security

Food security problems are widely discussed in the context of globally satisfying the demands of the growing population. This population and consumption growth will continue to increase global food demands for at least another 40 years (Godfrey et al., 2010). Related food security challenges include growing competition for land, water, and energy and reconciling these challenges with temperature changes and their impacts on agricultural productivity. Particularly, food producers are experiencing greater competition for land, water, and energy, with many negative impacts on food production. These challenges require changing the way food is produced, stored, processed, distributed, and accessed. Some of the core challenges to worldwide food security are highlighted in the following sections.

4.10.1. Soil Quality and Land Use Policy

Soil and land quality have significant impacts on food security. In addition, the current food production, storage, transportation, distribution, and marketplace environments are important. Globally, greater soil carbon contents offer greater mitigation potential. Another challenge is restoring wastelands and degrading soils and ecosystems. Innovative farm management practices could be used as methods for restoring soil organic carbon contents. Sustainable agricultural management practices can improve food security and soil quality simultaneously, which is exemplified by sustainable intensification (Royal Society of London, 2009). However, to feed a growing population, modern low-impact agriculture, which is also termed precision agriculture for site-specific management, could ensure soil quality over time.

4.10.2. Irrigation and Water Management

Improved water management is important for meeting the water requirements of rain-fed agriculture. Accelerating aridity and variable rain events can require frequent irrigation; however, irrigated agricultural systems could serve as a much more responsive and versatile approach. Adjusting large-scale irrigation schemes will be important for increasing ultimate production to correspond with demand, and improved water management practices are required. The water market is a looming issue in water-scarce regions, such as Asia and Africa, which lack the physical and economic resources for staving off a water crisis.

4.10.3. Food Distribution and Allocation

Sustainable agriculture also includes global food price patterns as indicators of food availability, at least for those who can afford it and have access to world markets (Godfrey et al., 2010). Food needs are not only related to adequate food availability but also closely to people's access to food. Access to food supplies can be maintained through regional distribution systems that focus on food health, locality, and quality. Previously the primary solution to food shortages was to devote more land to agriculture and exploit other natural resources. Crop and livestock productivity vary widely with geography, even across regions with similar climates. The best yields that can be obtained locally depend on to the abilities of the farmers to access and use seeds, water, nutrients, pest management tools, soils, biodiversity, and knowledge, among other factors. Important methods exist for producing more food in the same area of land while reducing environmental impacts. This method has been called 'sustainable intensification' and increases food yields by using existing technologies.

4.10.4. Increasing Production and Reducing Post-Harvest Loss

Increasing food production is another looming challenge in food security. The FAO predicted that it is possible to increase food production by 60% before 2050. This assumption was made based on the prediction that food production and yields from existing arable lands would increase due to the implementation of agricultural research (Konuma, 2014). Post-harvest loss is another challenge that must be minimized in developing countries. Nearly 14–16% of post-harvest losses occur in developing countries, beginning with the supply chain production, transportation, and processing stages. Another method for increasing production is to expand biotechnology that can produce high-quality seeds with immunity to diseases and infection during and after production and harvest. These new technologies can potentially increase sustainable production limits, which are based on new alliances of businesses, societies, organizations, and governments (Godfrey et al., 2010).

4.10.5. Reducing Food Waste

Food waste occurs at every post-harvest stage, including on the farm, during transport, processing, and retail, and in home and municipal food services. According to the data, approximately 30–40% of food is lost to waste in developed and developing countries for different reasons (Godfrey et al., 2010). In developed countries, most food loss occurs at the retail, food service and home stages rather than before the retail stage because retail services are always important. One survey showed that approximately 20–40% of fruits and vegetables are rejected in the UK before they reach stores because they do not meet the UK's high cosmetic standards. Even in some developed countries, expired and unwanted food is not used to feed animals or for any other purposes because of regulations aimed at preventing prion diseases. However, in developing countries, most food waste occurs because of poor food-chain infrastructures and lacking investments in or knowledge of food processing. In addition, these regions have limited food storage capacities. For example, one estimate is that the lack of availability of cold storage in wholesale and retail outlets results in wasting approximately 35–40% of fresh produce (Godfrey et al., 2010). Thus, different policies and strategies are required for minimizing food waste in developed and developing countries. Better infrastructure and investment in storage and education could help in developing countries. In contrast, awareness of high food prices and improved cosmetic standard policies could reduce food waste in developed countries.

4.10.6. Food Price

It is clear that food price trends affect food security and indicate food availability, especially for those who have the resources to purchase food and have access to it. Generally, because the poor spend more than half of their incomes on food, increases in the prices of staple foods force them to settle for reduced quality and quantity, leading to subsequent food insecurity and malnutrition. In contrast, high prices could encourage farmers to produce more food. However, farmers must also purchase the higher-priced food. Recent history reveals that the world experienced extreme price hikes for cereals and other major food commodities between 2007 and 2008, and 2010 and 2011. According to World Bank data, 44 million people transitioned into poverty due to increases in food prices between 2010 and 2011. Experts observed that this effect resulted from different phenomenon known as price volatility, which typically refers to highly unpredictable and remarkably large variations, especially price increases. These same experts emphasized that global food prices have become unstable and that this trend will continue in the future. The reasons for food price hikes and volatility are similar and correspond with the differences between supply and demand.

4.10.7. Changing Diets and Food Habits

Once major challenge to food systems is the rapidly increasing demand for staple foods, such as rice, meat and dairy products. In developing countries, meat represents one of the most concentrated sources of vitamins and minerals. However, much of the grassland used for producing rice to feed these animals cannot be converted to arable land, could only be converted with large, adverse environmental impacts. Reducing the consumption of meat and increasing the proportion of nutrients that are derived from more efficient sources would result in the ability to feed more people. Well-balanced diets that are rich in grains and vegetables and based on local food products are considered healthier. Local food consumption is commonly associated with efforts to improve food sufficiency and food diversification, particularly at the household level.

4.10.8. Population Growth

Population growth is a prime barrier preventing the achievement of food security, and it will become even more difficult to feed everyone in the future. The FAO by 2050 and 10.9 billion by 2100. The projected population growth rates are highest for developing countries (e.g., a 39% increase by 2050 and an additional 17% by 2100). The most severe growth will occur in the world's least developed countries, and demands for food are expected to double by 2030. This projection clearly requires greater food production, an obvious challenge given the world's limited arable land, which is decreasing in many countries due to urbanization, environmental degradation, and climate change. In addition, food distribution strategies must be improved to increase food access. This scenario is very similar to that of coastal areas, where excess population growth is affecting food security. In these areas, people are highly dependent on fisheries for protein and income. Recent research shows that fisheries that previously provided 80% of the dietary protein of local residents are now decreasing because of overpopulation while fish is increasing (Castro and D'Agnes, 2008).

4.10.9. Agricultural Biodiversity

Agricultural biodiversity is an important component for enhancing production by increasing resilience to environmental conditions and stress (drought, salinity, flooding). Climate alterations can influence the health of forests by increasing forest fires, blighters, and diseases. Ecosystem characteristics and activities (such as genetic resources and food cycling) are critical measures of resilience and highlight the need to mitigate agribusiness processes, which will be increasingly important with climate change. In addition, biodiversity in aquatic products (mainly fish, aquatic Mollusca, and crustaceans) will be important. In many regions, aquaculture has been profitable for food security, especially in archipelago regions. Future gains may also result from better stock selection with larger-scale production technologies. Gains in sustainability could result from concentrating on lower trophic-level species and from both aquatic and terrestrial food production. In addition, it will be important to use a more strategic approach for coastal zone management that emphasizes site location and capacity within catchments.

4.10.10. Adaptations to Climate Change

Climate change has an enormous impact on the agricultural sector and could affect food security. Because of climate change, optimal farming conditions are less frequent and crop areas have changed. Furthermore, crop productivity has decreased, the risks of agricultural pests, diseases, and weeds have increased, and crop quality has decreased. Meanwhile, the risks of floods, drought and water scarcity are increasing. Climate change also affects water quality and could result in soil erosion, salinization, desertification and rising sea levels, which are the most common impacts of climate change. However, different agro-climatic zones experience different impacts. Apart from these common impacts, sudden disasters could occur that would affect local agriculture. Thus, the adaptation process and options would vary from case to case. For example, aerial and satellite imagery can be used to monitor plant mortality and identify the invasive plant species that harm plants, which could help farmers adjust their agricultural processes. Financial support and knowledge of modern techniques could be used to collect rainwater, conserve soil moisture and efficiently use water. The loss of livestock due to extreme heat could also be reduced by increasing shade and improving ventilation in barns. Furthermore, farmers could use buffers and adjust their fertilizer and pesticide use during heavy rainfalls. Many approaches could be used to successfully address climate change, including planning for climate change and variability. For example, national climate change monitoring systems could include seasonal and annual predictions, decentralize community-run early warning systems, and provide climate insurance. For soil nutrient management, several common approaches could be considered, such as slow-forming terraces, conservation tillage, and integrated soil nutrient management. For sustainable crop management, crop diversification using new varieties developed through biotechnology, ecological pest management and seed and grain storage are effective (Zhu et al., 2011).

Our climate is evolving, and given the amount of greenhouse gases currently in our atmosphere, these changes will continue. Extreme weather events, such as high temperatures, droughts and floods, will become more frequent and severe, and will have tremendous social, economic, and ecological consequences. In the coming decades, world climate change could have an overall adverse effect on agricultural production and bring us towards, or perhaps over, critical thresholds in various regions. The sustainability of food production and feeding the overburdened population with limited natural resources will be one of mankind's greatest challenges. Systematic policy planning will be necessary for identifying relevant food production indicators and stakeholders from environmental and ecological perspectives. The project perspective has been considered as more reductionist and is dominated by individual ideas and philosophies. Climate change and food security have become global concerns. Human migration from food-insecure to food-secure regions will continue, with the consequences of overburdening limited resources and greatly influencing the quality of urban life. In the coming decades, systematic analysis is needed at the domestic, regional and international levels to incorporate the perspectives of multiple stakeholders, including donors, beneficiaries, and consultants. In the following chapter, we aim to contribute the key points of this systematic analysis by identifying the key domestic indicators and stakeholders.

Chapter 5

A SYSTEMATIC ANALYSIS OF SUSTAINABILITY: SUSTAINABILITY INDICATORS AND SYSTEM APPROACHES IN TRANSDISCIPLINARY PROJECTS

This chapter presents a systematic sustainability analysis by using indicators. The indicators are used to express sustainability and sustainable development and address human and environmental development indices. Sustainability is related to spatial and time scales. To work on a given time scale, a system analysis is helpful for understanding the behavior of complex systems over time. A system analysis also includes different indicator domains with affluent life.

5.1. PROJECTS AND SUSTAINABILITY INDICATORS

In this chapter, we explained how to develop projects by considering the specific use of sustainability indicators (SIs) by focusing on the following four major aspects: 1) the project scenario; 2) the stakeholder scenario; 3) accommodating multiple views; and 4) the system sustainability analysis idea. The four following domains are used for producing sustainability indicators 1) 'what is'; 2) 'what could be'; 3) project delivery; and 4) who is instituting sustainability as a project deliverable (Bell & Morse, 2008). These domains are used to measure how projects in the future could develop and evolve within the domain of scenario planning.

5.2. PROJECT SCENARIOS AND STAKEHOLDERS

When describing a project, a set of activities is defined for achieving outputs at the beginning and end of the monitored period to ensure that the system remains on the right track for its projects. To set up the projects, problem solving must be determined by using two types of project approaches, namely, 1) the traditional blueprint approach and 2) the process or 'adaptive' approach (Bell and Morse, 2008).

In the traditional blueprint approach, a cyclical relationship exists with the life cycle analysis. Traditional system development requires that system users consider proposed system plans by looking at detailed diagrams and detailed reports. These systems focus on efficient input into output processes and emphasize administration rather than management. The process or 'adaptive' project approach is used when inputs and activities have been used for making decisions during process management. This type of approach focuses on the realization of the project objectives rather than on the outputs. In addition, this project is intended to emphasize management rather than administration.

No project exists that is exactly right or wrong because each approach has strengths and weaknesses. The blueprint approach is specifically useful for control, accountability and administrative planning. This approach is good for situations in which the goals are clear and unambiguous and when the project objective is clearly formulated. The process-project approach is more flexible and can accommodate change and rethinking decisions (Bell & Morse, 2008). This type of project is likely best for situations in which the goals of the project can only be semi-formulated at their outset and when the project stakeholders are unclear regarding the delivery and expected outcome of the project. Each project comes with its own perspective and assumptions.

5.3. Transdisciplinary Scenarios in Participation

In 1994, Eyben and Ladbury defined participation as a process in which those with a legitimate interest in a project influence the decisions that affect them. In their opinion, participation is an important activity that results in reaching or failing to reach a goal. Hirschheim (1989) reported on several problems related with the success of the project. During the running time of the project, participants required the proactive inputs of consultants to maintain the experimental design. Some participants felt that they did not have time to dedicate to the process. Hirschheim was concerned about the procedure of formulating participation that was described by Craig and Porter for participation in the project context, including the element of control that creeps into a great deal of project work and explicitly describes itself as participatory. The authors proposed 'five practical suggestions' for those embarking on participatory project design, including 1) question and un-package the new participation orthodoxy; 2) advocate and influence; 3) make claims regarding efficacy and agency, 4) perform reflective analyses; and 5) make coalitions and negotiations (Craig and Porter, 1997).

5.4. A Transdisciplinary Approach for Sustainability

Participation and inclusion in the decision-making process have indicated problems to some people and have been discussed further. The participant stakeholders involved in a process project include the following: donors, project managers, implementers, and beneficiaries. However, because they are divided into subgroups, the complexity of the process increases. One reason for this increasing complexity and stakeholder participation problem is that the participant stakeholder coalitions in the project merge and vary with changes in the project context. The individuals and groups within and between coalitions will have various perceptions of project goals and purposes and may have significantly different views of what constitutes sustainability (Bell and Morse, 2008). The following diagram shows multiple views for project goals that lead to problems (Figure 5.1). This diagram indicates the vital importance of participation and dialogs between actors, which yield a new understanding. Therefore, we must link the system approach with participatory ideas. The following context shows the systemic sustainability analysis approach.

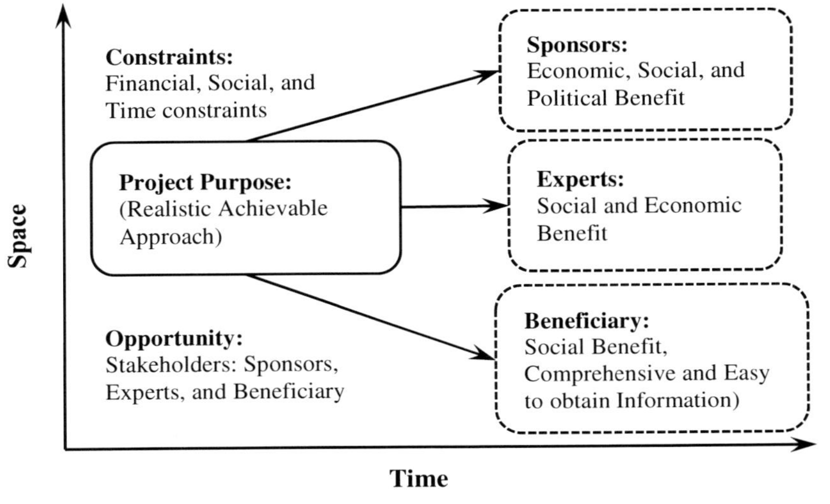

Figure 5.1. Multiple views regarding the purposes of projects on spatial and time scales.

5.5. Systemic Sustainability Analysis

In the system approach, after SIs have been conceived by using a reductionist approach that works without reducing complexity, the system focus is systematic and subjective (Bell and Morse, 2008). This approach should be prompt, participatory, qualitative, and descriptive. This method is called systemic sustainability analysis and is defined as the participatory deconstruction and negotiation of what sustainability means to a group of people. In addition, this method involves the identification and use of assessment indicators to assess this vision of sustainability. Sustainability analyses must be based on a number of assumptions. To develop a working model, of the following stages are required: 1) identify the stakeholders and establish the system to be measured. Because the fundamental concern is the identification of stakeholders, the system would likely refer to a task or primary issue related to overall sustainability at this stage and 2) identify the main SIs. The SI is subjective and depends on the stakeholder group. The main SIs must be drawn from a range of areas that reflect a holistic view and 3) identify bands of equilibrium (i.e., the reference condition). The stakeholders undertake this task, which empowers aspects of the process. The AMOEBA diagram is developed to represent multiple SIs in one diagram to use for systems. The diagram is produced using systemic, scenario-making, and participative approaches. The extension of the AMOEBA over time provides a snapshot indication of the sustainability of local food production management, as shown in Chapter 2. In all of the analyses, someone must lead an informed discussion and take action on items that are in disequilibrium. When imagining an approach, such as systematic sustainability analysis, the approach should not be considered as the only approach that could be used to introduce multiple perspectives to a sustainability analysis. No single definition describes sustainability, and its measurement remains subjective.

5.6. Project Outcomes and Sustainability

Although systemic sustainability analysis is good because lower and upper limits exist for the expectations of a stakeholder group, it can be problematic because of the lower limit discussions among stakeholders that are needed to arrive at a sustainable figure. A panel of experts is appointed to devise a figure that stakeholder groups may or may not agree with. In the participatory approach, an expectation can be made for the non-participatory alternative.

However, this approach remains problematic because participation does not mean that participant voices are heard and included to form policy change. The top-down vision for achieving sustainability is a worse idea. The upper limit is a more serious problem because it is required to reach agreements. A range of possible actions exists, from gaining insights into the preferences of social groups to setting a reasonable limit for expectations. Moreover, the discussion of the upper limit is sensitive, and thus no open discussion occurs in public forums, except on some limited websites.

Sustainability indicators based on participatory approaches address the issue of managing expectations. Although the projects and expectations are closely linked, they are contradictory. The projects are contained, limited, and time-lined, the risk is assessed, and expectations regarding sustainability are unconstrained, unlimited, eternal and without assessable cost. In this sense, achieving sustainability is a challenging task. An interesting observation is the way in which the use and development of SIs are themselves indicators of the very heart of the sustainability debate. Some opinions regarding sustainability refer to the environment, but others argue that sustainability does not have a single meaning. Success is considered as having the right knowledge and someone in place to receive it.

Sustainability considers what we want the future to become. If we simply implement what we think sustainability is, it may change our vision of what sustainability really means. To cope with this problem, the best method is to balance people's opinions of sustainability when an indicator is implemented. Although they are not precise measures, indicators can provide power and compromise between expectations and what is achievable without causing harm. Actually, all people are creators of sustainability. Although we want to achieve an unchanging set of sustainability indicators, we must meet the challenge of keeping pace with people's conceptions, ideas and dreams and try to make them real.

5.7. Conflicts of Interest and Mind Set

Notions of different worldviews and paradigms in sustainability have been established in the literature. The problem of sustainability indicators is the inability of different agencies and individuals from different cultures to implicitly or explicitly expose their different worldviews and assumptions regarding how the world works (Bell and Morse, 2008). In the future, it will be important to develop a vision of cultural change and to incorporate new ways of thinking into our analysis. This is a complex and problematic area because we tend to view our world from our own perspective and see our own prejudices and preconceptions.

The most difficult phase of systemic sustainability analysis theory is the transition from discussion to action. Talking must result in action and individuals may need to change their behavior in accordance with their experiences as part of an analysis. It is important that the outcomes should be implement accordingly. Therefore, the methodology in this chapter is not circular or concentrated but is an approach to collective activism. The use of sustainable indicators within policy contexts has received remarkable attention in the sustainability literature. However, wider studies have considered how policy is generally influenced. Sustainable indicators function as an essential element in evidence-based practice systems for reflection. Much of the available literature appears to focus on theories or opinions that form an affirmative system.

5.8. The System Dynamic Approach

Through the action of the learning cycle, a systemic approach to problem-solving can be applied in a systemic manner, expanding on ideas and cybernetics. Numerous systemic approaches are available for problem solving. However, sometimes approaches are systemic in the sense of recognizing and developing concepts of multiple views relative to reality. Currently, a psychoanalytic approach called the double task is being applied as a method for allowing a group of stakeholders to develop deeper and more subconscious concerns regarding their understanding of sustainability. Bell and Morse mentioned using the imagine approach in other contexts to obtain better explanations and applications. In addition, these authors mentioned how to develop a systemic analysis while creating an approach that has elements of imagining that allow for a greater sense of understanding regarding why differences in perspective can occur rather than accommodating them as a prelude to "use".

Questions remain regarding how to prepare an SI assessment in a holistic fashion. In describing the history of the information system of thinking, Checkland and Holwell (1998) described the area as 'the anatomy of confusion'. Furthermore, although SI could be confusing at some point, the confusion would essentially lie in blending technocratic and organizational mindsets. With SIs, a frightening mixture of mindsets can be observed, in which the technocratic element dominates. One example of a system model for integrating all economic and ecological indicators for sustainable development is available (Figure 5.2). The primary target of this model is to provide sustainable development, which addresses several indicators. First, the economic indicators consisted of income, consumption, and investment. Second, the environmental indicators consisted of environmental conservation and the development indicators were indicated as affluence in life. Several loops are shown in this figure. The income loop (income-economic indicators-income, R1) can be observed in which economic indicators become better as income increases. When the economic indicators increase, income also increases. In the consumption loop (consumption-economic indicators-consumption, R2), the economic indicators improve if consumption increases. In addition, the economic indicators increase consumption. In the investment loop (investment-economic indicators-investment, R3), if investment increases, the economic indicators increase as investment increases. In the affluent life loop (economic indicators-income-affluent life-consumption-economic indicators, R4), if income increases, the standard of living increases. Consequently, consumption increases and the economic indicators improve. In the technological strength loop (economic indicators-investment-technological strength-affluent life-consumption-economic indicators, R5), the technological strength improves if investment increases because R&D investment increases. Consequently, the standard of living is higher, consumption increases, and then economic indicators improve. In an affluent life (affluent life-development indicators-affluent life, R6), the developmental indicators increase when life becomes more affluent. The developmental indicators also increase and result in better life. In the environmental conservation loop (environmental conservation-environmental indicators-environmental conservation, R7), if environmental conservation increases, the economic indicators improve and the economic indicators increase.

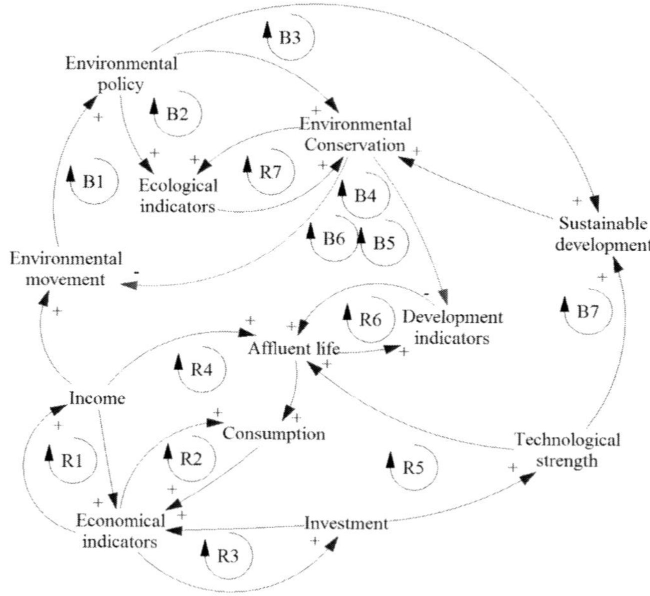

Figure 5.2. System approach for integrating economic and environmental indicators for sustainable development.

In this section, the effects are referred to as environmental conservation (environmental movement-environmental policy-environmental conservation-environmental movement loop, B1). Environmental indicators are required to determine environmental movement (environmental movement-environmental policy-environmental indicators-environmental conservation-environmental movement, B2). If the environmental movement increases, the policies will grow, which would result in better environmental conservation. However, better environmental conservation will decrease the environmental movement. The effects of environmental policy on sustainable development are common in the policy loop (environmental movement-environmental policy-sustainable development-environmental conservation-environmental movement, B3). Environmental policy controls sustainable progress. If sustainable development is better, environmental conservation improves and the environmental movement decreases. Indicators and affluence in life correspond with income, consumption and other indicators (Income-economic indicators-environmental policy-environmental indicators-environmental conservation development indicators-affluent life-consumption-economic indicators-income, B4, B5, and B6). If environmental conservation improves, the development indicators decrease because industrial development may be small in terms of environmental conservation. Moreover, if the income decreases, the environmental movement may decrease, as shown by the relationship between developing and advanced countries. Technological strength affects sustainable development, (technological strength-sustainable development-environmental conservation-development indicators-affluent life-consumption-economic indicators-investment-technological strength, B7). Sustainable development will improve if the technological strength improves.

5.9. Simplifying Complexity Using Indicators

The indicators that could be used to simplify complexity and importance have been highlighted in this section. For example, as individuals, humans would see the quality of education given to their children through the eyes of their subjective experience. However, those who are charged with providing resources increasingly see quality through the window of indicators. This experience would be partly based on experiences that include exposure to quantitative and qualitative indicators (the appearance of the classrooms, location of school etc.). All of this information is subconsciously processed in an integrated or aggregated fashion in our brains. Indeed, indicators are increasingly being viewed as a tool within a broader policy for facilitating "informed choice" by a much wider group than simply policy makers and managers.

Indicators are usually presented in "league table" format. This format can be used to compare one school with another by simply keeping an eye on their relative positions within the table. The league table shows listings that compare sports teams or individuals, institutions, nations, companies, or other entities by ranking them in order of ability or achievement. A table or chart (such as a league table, a ladder or a leader board) may be used to display such listings. A league table may list several related statistics, but they are generally sorted by the primary statistic that determines the rankings.

Many industries and institutions may compete in league tables to help bring in new customers and clients. A league table is an essential part of the competitive process, and a balance of incentives (positive or negative) has been designed. Quantitative and qualitative indicators are tools that can be employed during development. When considering complexity and pressing needs, it is important to simplify procedures for operation. This generalization is true whether the process is top-down and based on modernization or bottom-up and locally led. Indicators can be used to simplify complexity and serve as a tool (Table 5.1).

5.10. A Composite Index Using a Mathematical Model

A composite index is always difficult to calculate. A systematic approach is required for handling composite indexes. Here, the recommended model decreased the number of indicators by combining them into a composite sustainable development index (I_{CSD}). Figure 5.3 shows a basic hierarchy used to incorporate indicators into the $I_{CSD.}$ A step by step calculation is performed using the following seven steps: (1) selecting, (2) grouping, (3) weighting, (4) judging, (5) normalizing indicators, (6) calculating sub-indices and (7) combining them into $I_{CSD.}$ The details are given in Table 5.1.

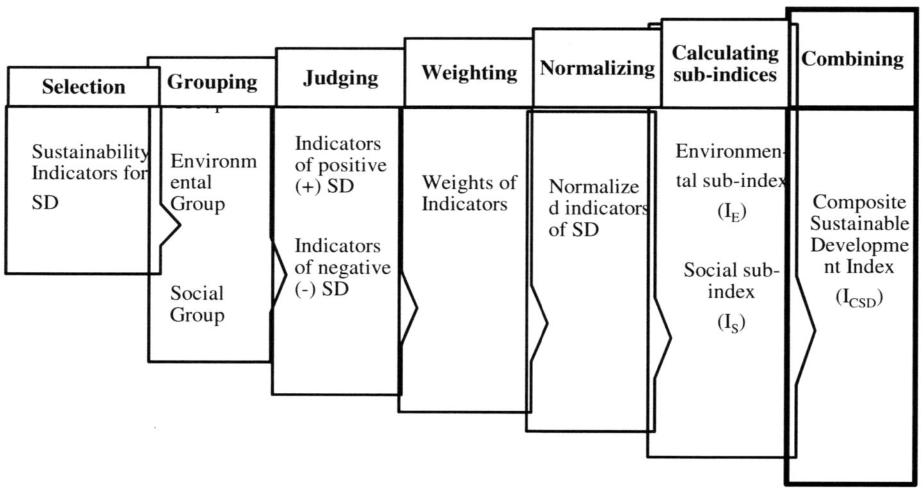

Figure 5.3. Calculating a composite sustainable development index.

Table 5.1. Methods for calculating composite indicators (CIs)

No	Method	Equation
1	Sum of country rankings	$CI_c^t=\sum_{i=1}^{N} Rank_{ic}^t$
2	Number of indicators above the mean minus the number below the mean	$CI_c^t=\sum_{i=1}^{N} sgn\,[\frac{x_{ic}^t}{x_{EUi}^t} - (1+\text{p})]$
3	Ratio or percentage differences from the mean	$CI_c^t=\frac{\sum_{i=1}^{N} w_i\, y_{ic}^t}{\sum_{i=1}^{N} w_i.}$, where $y_{ic}^t=\frac{x_{ic}^t}{x_{EUi}^t}$
4	Percentage of annual differences over consecutive years	$CI_c^t=\frac{\sum_{i=1}^{N} w_i\, y_{ic}^t}{\sum_{i=1}^{N} w_i.}$, where $y_{IC}^t=\frac{x_{ic}^t-x_{EUt}^t}{x_{ic}^t}$
5	Standardized values	$CI_c^t=\frac{\sum_{i=1}^{N} w_i\, y_{ic}^t}{\sum_{i=1}^{N} w_i.}$, where $y_{IC}^t=\frac{x_{ic}^t-x_{EUt}^t}{\sigma_{EUi}^t}$
6	Re-scaled values	$CI_c^t=\frac{\sum_{i=1}^{N} w_i\, y_{ic}^t}{\sum_{i=1}^{N} w_i.}$, where $y_{ic}^t=\frac{x_{ic}^t-\min(x_i^t)}{range(x_i^t)}$

Note: x_{ic}^t is the value of indicator *I* for country *c* at time *t*. w_i is the weight given to indicator *I* in the composite index. In method 2, *p* is an arbitrarily chosen threshold above and below the mean.

Indicators selection: First, appropriate performance indicators are selected to cover different elements of sustainability. The first preference should be the presentation of the indicators in a quantitative manner. However, some aspects of sustainability are more comprehensive and easier to understand using qualitative descriptions rather than quantitative descriptions. For example, social aspects are always presented in a qualitative way, but economic aspects should presented in a quantitative way. The first part of the selection process is paramount for comprehending the aspects of sustainable development for consideration. The selected indicators must be checked at regular intervals and should be assigned a symbol and unit of measurement. However, the ISO standard should be followed regarding the names and symbols of indicators.

Grouping selected indicators: Grouping is very important because it is strongly related to selected indicators. Grouping is performed according to the elementary aspects of sustainability (economic, j=1, environmental, j=2 and societal group of indicators, j=3). These groups represent the most accepted approach for defining sustainability (GRI, 2002). Here, the economic group of indicators considers the impacts of the company on the economic systems at the local, national and global levels. The indicators cover all aspects of economic interactions, including the traditional measures used in financial accounting (GRI, 2002). An environmental group of indicators covers the impacts of the company on living and non-living natural systems, including ecosystems and terrestrial, air, and water systems. A societal group of indicators reflects the attitude of the company towards the treatment of its own employees, supplies, contractors, and customers (IChemE, 2002).

Judging the Indicators: Indicators with increasing values that positively impact sustainable development are treated as positive performance indicators and are represented as $I^{+}_{A,ji}$, where i is the indicator of each group j. Simultaneously, a negative performance in sustainable development, could be presented by $I^{-}_{A,ji}$. For example, a greater soil pH value has a negative impact on soil fertility. However, greater crop production will have a positive impact on the economic performance of a country.

Weighting the Indicators: Weighting is the most difficult task in any context. A number of indicators are used to evaluate sustainability, which in turn are used to assess an organization as it moves towards sustainability. The importance of these indicators is very difficult to determine with sufficient accuracy (Afgan and Carvalho, 2004). It is challenging to determine the weights of indicators with insufficient data. Thus, a pairwise comparison technique is used to calculate the I_{CSD} and derive the relative weights of each indicator. The pairwise comparison technique is based on the method developed by Saaty (1995), which is called the Analytic Hierarchy Process (AHP). One example of the AHP model was presented by Krajnc and Glavic (2005). Here the AHP application was demonstrated for the sustainability assessment. In Chapter 2, we discussed local food logistics for multiple sustainability indicators. An AHP model for choosing criteria and subcriteria can be performed by using the Saaty Scale (Figure 5.4). The AHP employs a fundamental scale of absolute numbers to express individual preferences or judgments.

We have assumed that N indicators of sustainable development are being considered to quantify the relative weights of each indicator with respect to all other indicators of group j. This task is performed using pair-wise comparisons between the pairs of indicators. Comparisons are made by posing a question in which the two indicators (i and k) are more important for the SD of the project. The intensity of preference is expressed on a factor scale from 1 to 9 (Hafeez et al., 2002). A value of 1 indicates equality between the two indicators and a preference of 9 indicates that one indicator is 9 times more important than the one it is being compared with. This scale is chosen because comparisons are made in this way over a limited range in which the perceptions are sensitive enough to make distinctions.

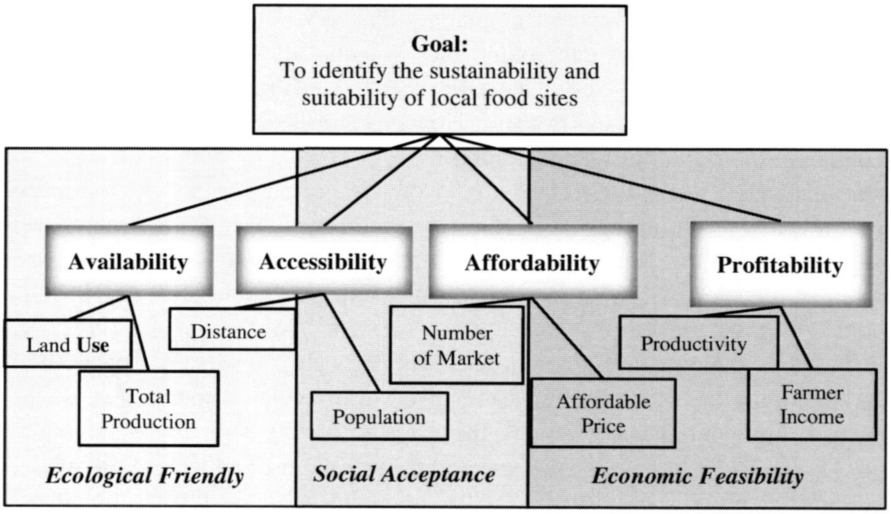

Figure 5.4. The AHP model for identifying and prioritizing local food production sites.

These pair-wise comparisons result in a (*NxN*) positive reciprocal matrix of *A*, where the diagonal is $\boldsymbol{a_{ik}=1}$ and the reciprocal property is $a_{ki} = (1/a_{ik})$, *i*, *k*=1,….,*n*. Assuming indicator ***i*** is 'p-times' the importance of indicator ***k***, then, indicator ***k*** is '1/*p*-times' the importance of indicator ***i***. A quick way to find the normalized weights of each indicator is to normalize each column in matrix *A* (dividing an indicator's relative weight by the sum of the relative weights in the column) and then average the values across the rows. Next, this average column represents the normalized weight vector of the containing weights ($\boldsymbol{W_{ji}}$) of the selected sustainability indicators.

Normalizing the indicators: Different indicators may be presented in different units. Thus, it is difficult to combine the indicators into I_{CSD}. Equations (1) and (2) are used to normalize the indicators as follows:

$$I^{+}_{N,ijt} = \frac{I^{+}_{A,ijt} - I^{+}_{min,jt}}{I^{+}_{max,jt} - I^{+}_{min,jt}} \tag{5-1}$$

$$I^{-}_{N,ijt} = 1 - \frac{I^{-}_{A,ijt} - I^{-}_{min,jt}}{I^{-}_{max,jt} - I^{-}_{min,jt}} \tag{5-2}$$

Where $\boldsymbol{I^{+}_{N,ijt}}$ is the normalized indicator ***i*** of the "more is better" type for the group of indicators, ***j***, for time (year) *t*. In addition, $\boldsymbol{I^{-}_{N,ijt}}$ is the normalized indicator ***i*** of the "less is better" type for the group of indicators ***j*** for the same time (year) t. Thus, the probability of incorporating different types of quantities with different units of measurement (i.e., physical, economic) is offered. One advantage of the proposed indicator normalization is the clear compatibility of different indicators because all indicators are normalized.

Calculating the sub-indices: The I_{CSD} calculation is a step-by-step procedure of grouping various basic indicators into sustainability indicators of ***j***. The sub-indices can be derived as follows:

$$I_{S,jt} = \sum_{jit}^{n} W_{jt} I_{N,jit}^{+} + \sum_{jit}^{n} W_{ji} I_{N,jit}^{-} \quad (5\text{-}3)$$

$$\sum_{ji}^{n} W_{ji} = 1, W_{ji} \geq 0 \quad (5\text{-}4)$$

where $I_{S,jt}$ is the sustainability sub-index for a group of indicators ***j*** in time (year) ***t***, and W_{ji} is the weight of indicator *i* for the group of sustainability indicators ***j*** and reflects the importance of this indicator in the sustainability assessment of the project.

Combining the sub-indices into the I_{CSD}: Finally, the sustainability sub-indices are combined into the composite sustainable development index, I_{CSD}, as follows:

$$I_{CSD,t} = \sum_{jt}^{n} W_j \,.\, I_{S,jt} \quad (5\text{-}5)$$

where W_j denotes the factor representing the a priori weight given to group *j* of the sustainable development indicators. These weights should reflect the priorities based on the opinions of the decision makers. In the final I_{CSD} calculation, an approach that uses estimated weights can be considered. These weights reflect the importance given to the economic, environmental, and societal performance of the project. Several indices and ratings cover various areas. Some of them are important from a subjective point of view and are described in the following section.

5.11. Indices Used for Innovation, Knowledge, and Technology

The purpose of this indicator is to measure a country's capacity to create knowledge by summarizing different indicators of national investment in highly qualified human resources in science, technology, research and education. All sub-indicators are measured per capita to counteract the effects of the sizes of the countries (Singh et al., 2009). Three basic components are combined between three and five underlying variables, which are mostly derived from the OECD database, to form the innovation index. The first component clusters variables such as basic research as a percentage of GDP and non-business researchers as a share of the labor force. The second basic component serves to build public-private links through data regarding R&D and the scientific contents of patents and publications. In the third component, data from business research, patents and the introduction of innovative products and processes are used to measure the innovative performance of the private sector.

The Innovation Index (SII) is part of the innovation scoreboard, which depicts achievements and trends, highlights the strengths and weaknesses of Member State performances, and examines European convergence in innovation (European Commission, 2001). Seventeen indicators from four major areas were used to analyze markets, namely; (1) human resources; (2) knowledge creation; (3) transmission and the application of new knowledge; and (4) innovation finance, output and the innovation scoreboard. References were taken between 1995/1997 and 1999/2000. The SII (of a given country) is equal to the number of indicators that are more than 20% above the overall mean EU, minus the numbers of indicators that are more than 20% below the overall mean EU. Here, the SII changes due to dissimilarities in the numbers of available indicators for each country. The index could vary between +10 (all indicators are above average) to -10 (all indicators are below average). Some indices are available that refer to the performance measure of the knowledge-based economy and convert the new knowledge into economic and technological progress to increase a country's competitiveness and the well-being of its citizens (Singh et al., 2009).

5.11.1. National Innovation Capacity

The core objective of the index is to create a quantitative benchmark for the national innovative capacity, which highlights the resource commitments and policy choices that most affect innovative output over the long run (Porter and Stern, 1999). The eight following sub-indicators were selected: (1) personnel employed in R&D, (2) expenditures on R&D, (3) openness to international trade and investment, (4) strength of protection for intellectual property, (5) share of GDP spent on secondary and tertiary education, (6) GDP per capita, (7) percentage of R&D funded by private industry and (8) percentage of R&D performed by universities.

5.11.2. Technology Achievement Index

This index is designed to determine the performances of countries for creating and diffusing technology and for building a human skill base (UNDP, 2001). The index considers data from 8 indicators that are clustered in four areas. First, technology creation as measured by the number of patents granted to residents per capita and by the receipts of royalties and license fees from abroad per capita. Second, the diffusion of recent innovations is measured in terms of the number of Internet hosts per capita and the share of high and medium technology exports among total goods exports. Third, the diffusion of old innovations is measured based on telephones per capita and electricity consumption per capita. Fourth, human skills are measured in terms of mean years of schooling in the population aged 15 and above and the gross tertiary science enrollment ratio. In the ICT, the index attempts to provide a general feature of a country's situation in terms of development and the application of information and communication technologies (Fagerberg, 2001). Five simple indicators are used, including the number of mobile telephones and the number of Internet users that are used as components for developing the composite indicator.

5.11.3. An Index of Sustainable and Economic Welfare (ISEW)

The Center for Environmental Strategy (CES) and New Economics Foundation (NEF) developed the ISEW. This index represents one of the most advanced attempts to design an indicator for economic welfare. The basic purpose of this index is to evaluate economic activity, which is performed for the welfare of people. To an extent, this index attempts to replace the GDP as an indicator of progress because the GDP is likely to provide wrong information because it does not distinguish between activities that improve or directly damage the quality of life (CES, 2000). Under this index, a group of 20 sub-indicators are used that include seven economic activities, including adjusted consumer expenditures, services from domestic labor, services from consumer durables, services from streets and highways, public expenditures on health and education, net capital growth, and net changes in international position, to address the welfare of people. In addition, the 13 following indicators are used to decrease welfare: consumer durables (the difference between expenditure and value of service), private expenditures on health and education, commuting costs, personal pollution control, automobile accidents, water pollution, air pollution, noise pollution, loss of natural habitat, loss of farmlands, depletion of non-renewable resources, the costs of climate change and the costs of ozone depletion (Guenno and Tizzi, 1998). The European Commission (EC) developed an index for measuring regional problems in the community that is known as the "relative intensity of regional problems in the community" (Commission of the European Communities, 1984). The purpose of this index is to help the community acquire regional policy that emphases reinforcing the economic performance of regions that are experiencing development delays. The commission has implemented the following three sub-indicators to evaluate the relative intensities of regional problems at a community level in a global and synthetic way: (1) the GDP per employee in the ECU, (2) the GDP per head in the PPS, and (3) the unemployment rate.

5.12. Development Indicators

The most well-known theories of indicator development that were employed to gauge the economic performance of a country are encompassed in the Gross Domestic Product (GDP). The GDP is one of a suite of indicators, including the GNP, that account for various aspects of an economy, such as the consumption of fixed capital. However, the GDP has several limitations in a growing economy. For example, GDP growth sometimes is incurring significant damage to its environment and natural resources while numbers of industry expand throughout a country or a region. Even if the damage is aesthetically or morally defensible at some stage, there would be a need for realignment. The GDP and GNP-related discussion was made in Chapter 4. However, poverty is obviously an important element to consider in advance and it is the 'goal of development' to emphasize reductions in poverty as a primary measure of success. However, poverty means very different things to different people. Thus, this subject is a very complicated problem. By using a commonly applied approach, Albert highly simplified this method to consider the notion of 'poverty line'. This is the indicator of an income level at which people are not in a position to afford subsistence or basic needs. Some common terms are used to note more details regarding the poverty line.

5.12.1. Human Development Index (HDI)

This widely used index was developed by the United Nations and attempts to measure three core aspects of human development. The three aspects include a long and healthy life, knowledge and GDP per capita (UNDP, 2001). A long and healthy life is measured based on life expectancy at birth. Knowledge is evaluated based on the adult literacy rate (two-thirds weight) and the combined primary, secondary and tertiary gross enrolment ratio and GDP per capita (Purchasing Power Parity, PPP, US$). All of the indices are measured based on the minimum and maximum values for each indicator, and the performance in each dimension is expressed as a value between 0 and 1 (Morse, 2004).

5.12.2. Headcount Ratio (H)

The headcount ratio is the simplest indicator that indicates the proportion of the population below the poverty line. The headcount ratio is symbolized by H (the number of poor divided by the total population). The headcount ratio obviously depends on the level of the poverty line. Of course, if is the headcount is adjusted too low, it becomes meaningless. The primary problem with the headcount ratio is that it does not account for the depth of poverty. Instead, we know how many people are below the poverty line but cannot know how far below the poverty line they are.

5.12.3. Income Gap Ratio

The income gap ratio focuses on the depth of poverty below the poverty line. This indicator represents the average deviation from the poverty line for people who earn less than the poverty line. Therefore, it does not include those who are not in poverty. Deviation relative to the poverty line is equal to $\frac{z-y_i}{z}$, where z is the poverty line and y_i is the income of person i. Summing this up for all individuals below the poverty line and dividing it by the number of people results in an average deviation from the poverty line.

5.12.4. Poverty Gap Index

The Poverty Gap Index is one method for handing the lack of sensitivity in the income gap ratio to the proportion of people below the poverty line through modifications by using the headcount ratio. The poverty gap ratio weights the value of the income gap ratio by considering the population classified as poor in terms of the poverty line. However, a problem exists regarding this indicator, and the poverty gap index does not cover the nature of distribution for the income of poor people.

5.12.5. Index of Dissimilarity

Inequality is an important factor within development. A number of measures of equal distributions of income amongst individuals within a population exist. The simplest way to approach this measure is to rank the population in terms associated with the category. The differences between the proportions of persons in a category and the proportion of wealth in the same category will provide some ideas regarding the extent of the quality. For example, the formulation of the Index of Dissimilarity (ID), which is shown as follows:

$$ID = 0.5\sum |Xi - Yi| \quad (5\text{-}6)$$

When X is the income category width and *Y* is the proportion of income associated with that category width. Higher *ID* values correspond with greater inequality. The best known and most widely used measure of equality is the Gini coefficient. However, the Gini coefficient is a one-dimensional measurement in which only inequality in income is covered.

5.12.6. Sen's Index of Poverty

Sen's Index of Poverty (*P*) is a combination of the headcount ratio, the income gap ratio, and the Gini coefficient for the population classified as poor.

$$P= HI+ (1\text{-}I)\, Gp \quad (5\text{-}7)$$

Here, *H* is the Headcount Ratio, *I* is the Income Gap Ratio, and *GP* is the Gini Coefficient below the poverty line.

5.13. Integrating Development Indicators

A small number of examples of aggregated development indices originated from the work of the UNDP. Composite indices reduce complexity and make it easier to understand. However, the simplification process carries the potential danger of hiding information that may be useful, or at least should be considered during interpretation. In Sen's Poverty Index, it is important to present all of the components to be interpreted properly. This is also necessary for HDI and its relatives. Identical values for two or more countries can be obtained that have very different balances regarding the components of indices, such as the HDI, in addition to the notes on the methodology employed. However, the user may not want to know how an index is calculated or the complexities involved in the process.

The HDI and its relative transparency are calculated because only three equally weighted components contribute to its broad acceptance (Streeten, 1994). The indicator can be presented as a "league table" presentation of countries that generates the 'name and shame' style and has undoubtedly played a major role in enhancing its popularity. The HDI has helped shift the reporting agenda away from narrow economic indices and provides focus for media to report and debate within the practitioner and academic communities. Furthermore, the HDI has encouraged a host of constructive debates regarding the nature of human development and how to best gauge progress. The HDI and other indicators are by no means the only such aggregated indices of development. However, rather than simply cataloging more examples, two indices were selected for further discussion in an attempt to gauge the elements that are now considered vital for development. First, as its name suggests, ESI is an index intended to gauge the environmental sustainability of countries. Yale and Columbia University created the ESI, which is intended to estimate environmental performance. The criteria of this index are discussed above in Chapter 2. Another complex index is the Corruption Perception Index (CPI), which was created by Transparency International (TI). The CPI aims to reflect an important aspect of "good governance", a feature that is seen as an important requisite for development. The CPI is designed to proxy corruption. This index is distinguished from the ESI based on the perceptions that the individuals giving bribes for different purposes in a country. The data sets have more uniformly constructed CPIs, and the steps involved in obtaining the index are not very complex.

In human development index, data limitations can completely prevent measurement attempts. This index was made to aid management rather than to help people understand the nature of human development. Indeed, the HDI was purposely created to feed policy that is used by organizations to help prioritize interventions and measure success. As currently formulated, HDI risks are counterproductive and do not highlight the tremendous importance and improvidence of improvements in mortality and education for human development. The HDI gives the false impression that we know the levels of these essential activities in all developing countries. In some sense, the HDI devalues a host of efforts that are needed to obtain real and timely information on mortality and education in developing countries. For our purposes, we need to develop indices that are relevant to our specific conditions. All of the indicators (ecological and environmental) that relate to human development require a systematic analysis. Thus, the sustainability indicators could help to measure developmental progress over the period.

A systematic analysis using indicators for sustainability and system dynamics approaches is integrated to show the affluent life. In addition, human development indices need to be recognized to ensure that quality of life moves in the direction of sustainable development. To work with the time scale, the system dynamic approach is helpful for understanding the behavior of complex systems over time. Reinforcement for the planning and backcasting approach could be implemented to achieve the desired sustainability in the agriculture, environment, and renewable energy sectors. By viewing sustainability in a broad sense, we can compare and contrast application facts across these domains and apply lessons from one to another. In the following chapter, sustainability and change of thinking through backcasting approaches are discussed. In addition, the next chapter introduces and discusses alternative systemic approaches to thinking and problem solving by covering a wide variety of topics, such as changes in environmental ecology and renewable energy.

Chapter 6

Sustainability and System Approaches: Long-term Environmental, Energy and Ecological Modeling to Ensure Sustainability

This chapter reviews and discusses system-level approaches for thinking and problem solving by covering a wide range of topics, including the environment, ecology, and climate change. Although the recent issue of climate change is one of the most pressing environmental problems today, it is far from the only challenge. The millennium ecosystem assessment, a four-year international study on the status of the world's environment, reported that two-thirds of the globe's ecological services, such as the clean air and water provided by nature, are degraded or are used unsustainably (Millennium Ecosystem Assessment, 2005). A separate study conducted by academic institutions in five countries predicted that the sources of the world's natural resources are likely to collapse by midcentury due to overfishing, pollution and other environmental factors unless major changes are made. Therefore, we should be concerned about the unanticipated emergence of many contemporary environmental issues and expect new and unforeseen risks to suddenly appear. For example, poverty and other forms of social distress aggravate environmental degradation. In 2007, a study by the World Health Organization (WHO) acknowledged that approximately 40% of the deaths worldwide are caused by water, air, and soil pollution (WHO, 2007). Furthermore, the world population is approximately 6.5 billion and is continuing to increase, mainly in non-industrialized nations. In 1950, 57% of people were malnourished in non-industrialized nations and 20% were malnourished globally (2.5 billion). Human behavior is responsible for the design, construction, use, management and disposal of all appliances. The rapid growth of urbanization and the exponential growth of the world population are all problems. To balance all threats over time, a dynamic system approach and a change in thinking are required. Therefore, significant changes in behavior will be an important element for any successful effort to protect the climate and adopt a path towards sustainability. These transformations must increase economic prosperity, personal welfare and security in industrial and developing nations.

6.1. The Components of Sustainability

Sustainability provides visions and ideas for the future. A pathway to sustainability will be helpful for exploring the complex, dynamic interactions that occur among global, social, and human systems through transdisciplinary interactions. Whether the challenge is at the organizational or societal level, overcoming a culture's prevailing worldview is never easy. Every culture has a shared method for comprehending the world that is influenced by the different conceptual frameworks of its people. This is necessary for understanding events, deciding right from wrong, and shaping the ways in which people think about work, other people, and the natural environment. In contrast, technological advancements and innovations have been rigorously addressed in sustainability. Some technological innovations have resulted from curiosity-driven research, and other innovations have resulted from societal demands and policy intervention. Indeed, an increasing number of promising technologies that could help us address environmental problems and promote sustainable development have been invented through rigorous research at universities, research institutes, and corporations. Kajikawa (2008) highlighted the basic components of sustainability science, which include goal setting, indicator setting, indicator measurement, causal chain analysis, forecasting, backcasting, and problem-solution chain analysis. A number of studies have contributed to this method by addressing essential approaches and the promising tools for sustainability science, such as visionary and future scenario approaches, multilateral indicator systems, knowledge structuring, cause-effect analysis, and solution paths.

6.2. Sustainability and Solution Paths

New solutions always begin with a fresh perspective, and new visions and forms of reasoning must drive the transition. New perspectives begin with an accurate understanding of our current challenges and the beliefs, assumptions and values that produced the problems. In addition, new solutions require us to rethink our values, how we want to live, and what we want to achieve in the near future. In future studies, the dominant approach will still be a traditional forecasting approach. Forecasting studies require detailed research regarding the complex relationships among economic growth, development, demographic trends and access to resources, including nutrition (Arizpe et al., 1991). The results crucially hinge on the assumptions of future trends in these relationships and developments. Therefore, it is not surprising that predicted energy use differs by more than a factor of ten between different forecasts. However, it is possible to study the problem from another perspective, such as the backcasting approach. Based on prevailing trends, this approach is unlikely to generate solutions that would predict when these trends will break. However, discontinuities are likely to occur in the future and should be deliberately sought in some cases. Backcasting is recommended as an interesting alternative approach to understand the principles of sustainability, locate unsustainable processes, determine the changes required for improving the principles of sustainability, and form a vision of how to implement the suggested change.

6.3. Backcasting as an Alternative Approach

One major distinguishing characteristic of backcasting analysis is concern regarding how desirable futures can be attained rather than what futures are likely to occur. It is explicitly normative to work backwards from a particular desirable future end point to the present to determine the physical feasibility of that future and what policy measures would be required to reach that point (Robinson, 1982). The concept of backcasting was introduced in the 1970s by Lovins (1976), who used a 'backwards-looking-analysis' in his exploration of long-term energy policy in the United States. Assuming that future energy demands are primarily a function of current policy decisions, Lovins suggested that it would be beneficial to describe a desirable future (or a range of futures) and assess how such a future could be achieved instead of only focusing on likely futures. After the strategic objective is identified in a particular future, it is possible to work backwards to determine what policy measures should be implemented to guide the energy industry in its transformation towards that future.

In 1982, Robinson detailed this approach and developed a backcasting technique as an approach to traditional forecasting and designing strategies. This approach was notably designed to serve as a tool for exploring property policies. As described by Robinson (1982) and other authors, backcasting should support obtaining quick ethical and normative definitions from processes and explore desirable futures and scenarios rather than business-as-usual scenarios. The concept of planning desirable scenarios as a foundation for strategic action is not a recent one. Backcasting is primarily based on a philosophy and pragmatic critique of predictive forecasting, and this critique relies on the notion that science could be a human process that involves power and the exclusion of minority views.

Höjer (1998) outlined backcasting as a scenario technique that aimed to present solutions to issues that were not resolved according to typical scenarios, trends and forecasts. A related purpose is that predictive forecasting requires a closed system, whereas social developments over long periods cannot be categorized in these terms. Such future predictions are primarily based on assumptions regarding selection and behavior in the present and disregard the chance that new imperative issues could accommodate behavioral choices in the future. In epistemological reasoning, Dreborg (1996) argued regarding the essence of backcasting as an approach or paradigm rather than as merely a method. He explained that backcasting is related to discovery rather than justification. This idea is compatible with the notion that participatory methods such as backcasting should enable problem structuring rather than problem solving. If backcasting were examined from this viewpoint, it would be considered as a recipe instead of a tool, and as an overarching approach involving a range of specific strategies.

Steen and Akerman (1994) conducted other surveys of studies regarding the transitions of energy and transport systems in Sweden according to the Swedish Committee on Climate Change. One conclusion was that few studies have seriously accounted for sustainability. These studies are characterized by a vertical axis that represents whether a state can be regarded as in line with sustainable development or not (Figure 6.1). The shaded area reflects the differences of opinion regarding the sustainability criteria. This result indicates that today's society is generally not sustainable. One category (A-directional studies) investigates different economic measures and other measures that promote behavior based on the natural environment. The proposed measures will most likely work in the right direction. However, how close society will be brought to sustainability was not addressed in these studies. Other studies (B-short-term studies) have used short-term official goals as a starting point and have attempted to find a means of achieving them. These goals are typically only one step on the way to a real solution for sustainability. In addition, some studies apply a long-term perspective. The majority of these studies apply some type of forecasting methodology (C-forecasting studies). However, the predicted developments invariably do not respond with the sustainability goal, primarily because the adopted presuppositions are so restrictive.

Additionally, the chosen methodology precludes major change. A few long-term studies use a backcasting approach (D-long term studies) and begin with images of future states (scenarios) where the energy supply-demand satisfies the requirements of sustainability.

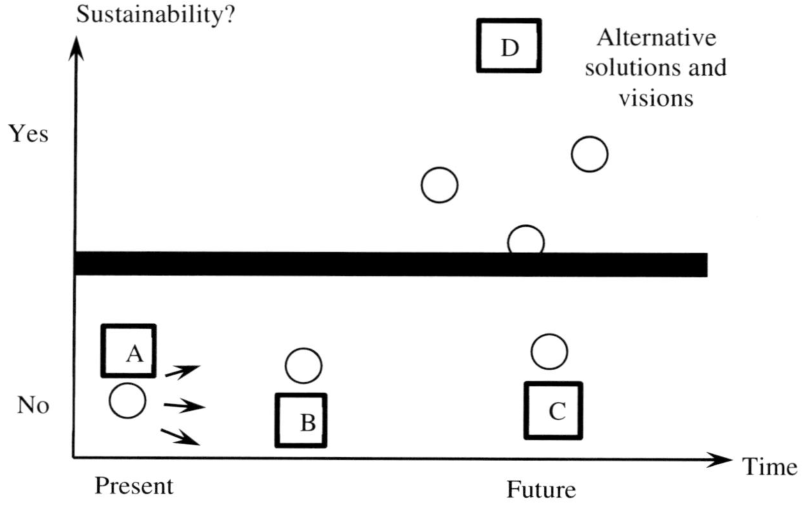

Figure 6.1. A. Directional studies, B. Short-term studies, C. Forecasting studies, and D. Backcasting studies (Steen and Ackerman, 1994).

The various approaches for operationalizing backcasting show similarities and differences. The characteristics that favor backcasting (as shown by Dreborg, 1996) are complex drawbacks that affect several sectors and levels of society, including an insufficient progressive approach, the need for major (systemic) changes, and the dominant trends portion of the problem. In addition, this problem is related to externalities that cannot be self-addressed by the market alone, and the time horizon is long enough (30 to 50 years) to permit for deliberate selection. One very important feature of this method is that it permits the analysis of different pictures of the long runs that are thoroughly analyzed regarding their feasibility and consequences. Every alternative must appear coherent, and the analysis of results for social life should be credible. Backdating should identify strategic choices for society to make all types of decisions. Moreover, Dreborg (1996) compared backcasting and forecasting based on philosophical issues. Forecasting has resulted in thoughts of causality, determinism, and justification, whereas backcasting has focused more on causality and teleology, partial indeterminacy, and the context of discovery. This perspective also focuses on dominant trends, likely futures, and possible marginal adjustments in forecasting. In addition, backcasting mostly addresses societal problems that require solutions, a desirable future, a scope for human choice, and strategic decisions.

Based on previous explanations, the starting point of the backcasting approach is the sustainable equitable and livable world at some point in the future. The requirements for this transition can be studied and compared to the present-day capabilities of stocks of material and human resources. Thus, backcasting does not aim to predict, but rather aims to discover constraints on the different paths that can be followed.

6.4. COMBINATIONS WITH BACKCASTING

Backcasting can be combined with a number of other tools to assess or promoting sustainability. Höjer and Mattson (2000) suggested combining backcasting with forecasting because forecasting indicates the need to step out of a scenario to shift a trend that may be part of the problem. Höjer (1998) combined backcasting and conventional Delphi in a survey. The Delphi technique is used for problem solving and decision-making. The following features are used in the Delphi method: the anonymity of responses from panel experts, synchronicity, controlled feedback, and statistical descriptions of responses. The objectives of most Delphi applications include the reliable and creative exploration of problems or the production of suitable information for decision-making. A combination of backcasting and the Delphi method has been used to reach convergence among experts regarding future scenarios by articulating the assumptions underlying different initial pathways. This combination could be useful in an interactive setting when a conventional Delphi is replaced by an interactive Policy Delphi. In a Policy Delphi, scientific experts, policy makers and other stakeholders are not allowed to converge but are used to generate the strongest possible opposing views on the potential solutions to a major policy problem.

6.5. BACKCASTING APPROACH FOR MODELING ENERGY AND ECOLOGY

Typically, backcasting is applied to long-term complex issues that involve many aspects of society and technological innovations and change. The focus of interest is on a perceived societal problem of great importance, such as the immense and growing effects of transportation on the environment (Dreborg, 1996). To fully understand the application of the backcasting approach, this chapter refers to the following book: "Transition to a Sustainable Society: A backcasting approach to modeling energy and ecology" (Mulder and Biesiot, 1998). This book provides current answers to energy and ecology issues based on a backcasting approach, which begins by defining a desired future situation (say in the year 2050) and results in the development of scenarios that connect the present situation with the end point situation. Such an approach produces insights into the challenges that lay ahead for realizing a future situation. The calculations regarding energy flows must realize household consumption levels, the needs of industrial systems, the amounts of natural resources consumed and the amounts of goods and services produced.

Backcasting works backward from the desired end point to the present. This is a normative scenario approach that is proposed for long-term changes of twenty to one hundred years and addresses changes in human behavior. This is one of the primary reasons that backcasting has gained greater attention in bioenergy and sustainability research. Bioenergies are integrated into technical and energy administrative procedures in outstanding positions due to their environmental sustainability. However, fossil fuel use is limited due to its production and effects on global warming as GHGs. One example of the backcasting approach is to use biomass energy resources to achieve environmental sustainability. In the conservation reserve program, land and marginal croplands are converted to perennial biomass energy crops, such as *Miscanthus*, sorghum or switch grass, to meet energy goals.

Improved plant materials and management practices are required for pasture and perennial biomass production systems to optimize productivity and profitability for sustainability. System management that focuses on natural resource conservation and environmental quality enrichment requires a thorough understanding of energy and the environment. As a first step, future goals aim to define desirable scenarios that recognize the energy potentials of crops, energy-exergy, bioenergetics and GHG emissions. Starting from the present state and projecting towards the future, several possible paths can be examined for reaching the desirable future scenario by using the backcasting approach (Figure 6.2).

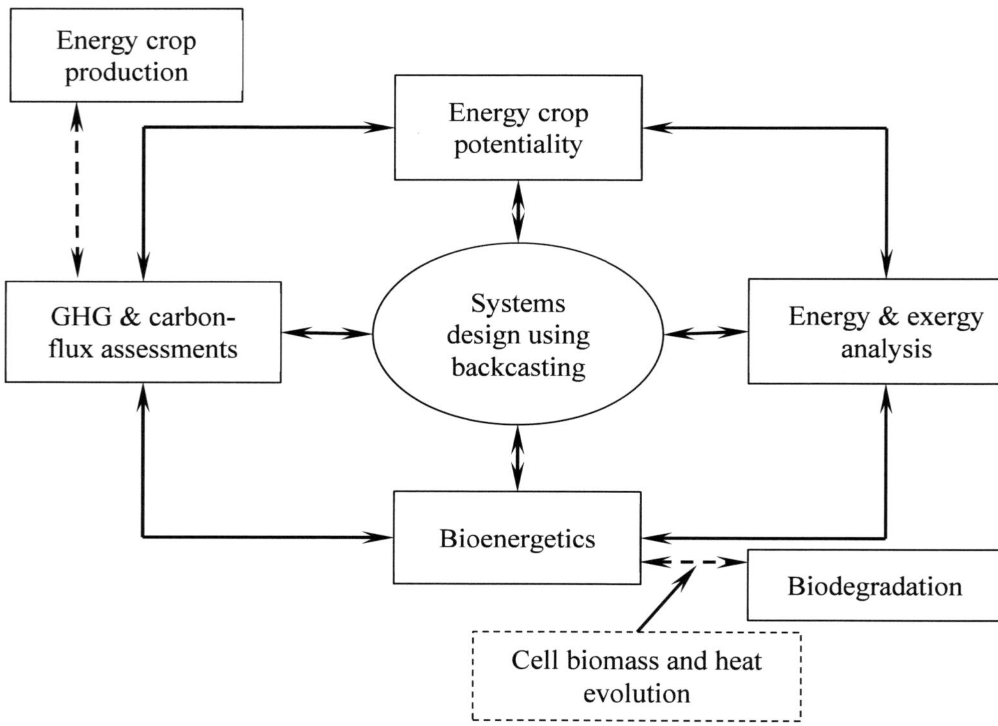

Figure 6.2. Design of the systems with different scenarios, including energy crop potentiality, energy and exergy analysis, bioenergetics and GHG and carbon flux assessments (Ting, 2009).

6.5.1. Energy Crop Potentiality

Energy crops can be used as biomass crops that simply provide high biomass outputs per hectare for low inputs and as specific products that can be translated into other biofuels, such as sugar or starch for bioethanol or vegetable oil for biodiesel. Biomass energy crops have energy potentials based on their geographical positions and on their sustainable use that are required for harvesting over time. The estimated potentiality and sustainable harvest of energy crops for feedstock could ensure sustainability over long periods by using the backcasting scenario analysis.

6.5.2. Energy and Exergy Analysis

Elemental analyses are necessary for determining the heat values of biomass. Growing biomass crops requires energy, and energy is required for converting harvested products to fuel. Energy crops, such as short-rotation woody crops, herbaceous woody crops, grasses, starch crops, sugar crops, forage crops, oilseed crops, switch grass, and *Miscanthus,* can be converted to liquid biofuels through thermochemical and biochemical conversion processes. To estimate energy and exergy relationships, this scenarios can be used to conduct elemental analyses and can result in higher heating values of feedstock production through energy and exergy analyses.

6.5.3. Bioenergetics

Bioenergetics is an energy relationship in living organisms. Microbial activity is a function of catabolic reactions for energy production and biomass synthesis. The relationships between energy and synthesis reactions are crucial for understanding the balance of inputs and outputs from the biodegradation process. Bioenergetics provides an analytical framework for combining the stoichiometries of these individual reactions with classical thermodynamics to estimate important results, such as cell biomass production and heat evolution. The resultant overall stoichiometric relationships are used to determine nutrient requirements, pH and alkalinity effects, and energy flow throughout the system. With bioenergetics, an analytical framework can be developed to understand the stoichiometry of cell biomass using classical thermodynamics and heat and mass transfer.

6.5.4. GHG and Carbon Flux Assessments

Bioenergy crops offset carbon dioxide emissions by converting atmospheric CO_2 to organic C in crop biomass and soil. However, their soils also emit nitrous oxide and have different effects on the oxidation of methane in soils. The soil carbon GHG flux trial showed the effects of bioenergy crop land-use changes on soil carbon stocks and GHG emissions. Soil carbon GHG flux assessments and mapping are needed to underpin recommendations for the environmental efficiencies of various bioenergy crops and their management. Thus, GHG and carbon flux assessments for environmental compatibility and sustainability are expected.

6.5.5. System Design Using Backcasting

Backcasting is an approach that identifies policies and programs for connecting a defined and desirable future to the present. In contrast with forecasting approaches that are used to predict future states, backcasting scenarios should be developed according to their desirability for achieving an optimistic and preferable bioenergy future (Figure 6.3). In system design, the integration of the above scenarios with a backcasting approach can bring the desirable future of energy to the present. Combining the knowledge of bioenergy use with existing energy resources in spatial planning decisions will be a primary challenge for long-term sustainable energy systems. Using backcasting approaches, the main research strategy will enable discussion and classification of analytical framework options without losing track of the more radical long-term transition goals of bioenergy to ensure environmental sustainability.

Because of a combination of a variety of potential developments in population growth and technology, a large number of backcasting scenarios have been evaluated in the models. These models are not based on narrow thought or rigid assumptions concerning future developments. The models find that a transition towards some type of sustainable future as defined above is physically possible but that the decision windows are rapidly closing. Only dedication to a much higher efficiency in the throughput of natural resources, a swifter transformation of the energy and materials supply system and a newer form of nonmaterial consumption will result in relatively undisturbed transitions.

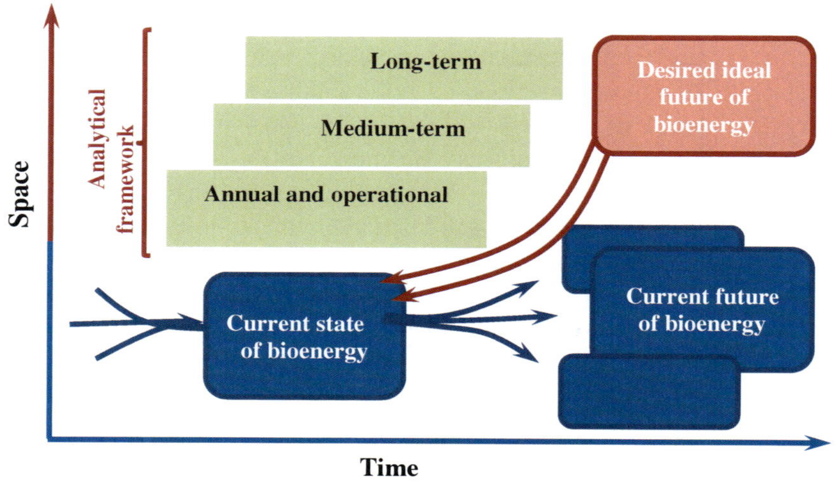

Figure 6.3. Backcasting from the desired future of bioenergy.

Mulder and Beisiot (1998) introduced the basic version of the transition calculation model of energy demand. The model is intended to investigate the physical possibilities of reaching a sustainable equilibrium world society at an aggregated macro-level. Constraints on the 'possible futures' are based on a limited amount of fossil fuels available for energy, and the use of this amount could be even more limited because of emissions into the environment. The transition-modeling framework begins by setting a target for sustainability in the future that uses energy and ecology in the model. A range is later specified for energy use within environmental (scarcity and pollution) constraints for this final, sustainable equilibrium. This model is shown in Figure 6.4, and the energy use in the typical situation varies between low and high assessments. From this range, some backcasting scenarios were devised from the end point to the present situation. These scenarios include combinations of population development, the development of service level per capita and the energy required per service (including equity), which dictates the position of the line representing the total energy demand. These scenarios also include assessments of the rates of renewable resource exploitation, which results in a line that represents supply from renewable resources.

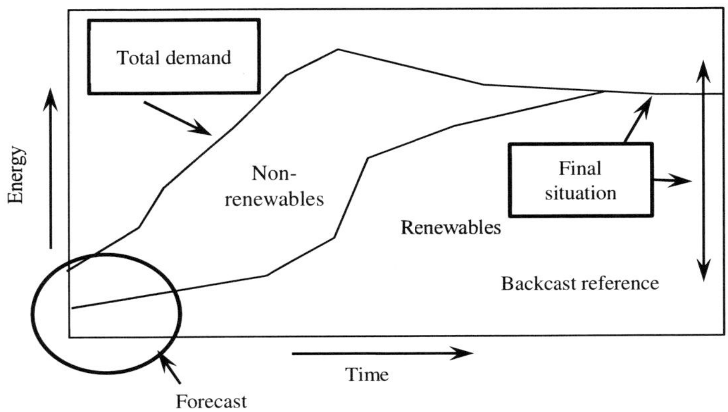

Figure 6.4. The transition to renewable energy sources and references for backcasting (Mulder and Biesiot, 1998).

These two constraints, the limitations of energy and emissions, dictate the size of the area between the supply line of the renewable energy source and the total demand. If only a few factors are initially assessed, the remaining factors can have a series of possible values. For example, estimates of population growth and the year in which equity is achieved can be combined with a chosen transition path to renewable energy. Because most forecast models only have limited time horizons (Figure 6.1), these models do not generally provide information regarding the potential long-term sustainability of their underlying assumptions. However, this scenario is based on a forecast of a low population, which allows for large-scale use of biomass without endangering the food supply. The Greenpeace fossil fuel free energy scenario is explained in a statement of conservation, which also contains the potential to reach equity and sustainability within the next century, with even higher population growth than assumed in the Goldemberg scenario. The Goldenberg scenario is briefly described in Mulder and Biesiot's book and claims that the end-use oriented global energy strategy is centered on energy-saving measures. The other scenarios (WCED, IIASA, and CPB) show the unsustainability of business-as-usual developments in various degrees.

The combinations for this model and the results indicate the existence of a sustainable development dilemma. If historic trends of growth in the service level and in energy efficiency improvement are continued and combined with equity (even including a transition to solar energy), it is implied that the depletion of fossil-fuel resources occurred to such an extent that greenhouse issues became more acute. The requirements for carbon-dioxide mitigation become almost insurmountable. Delays in changing the present trend will narrow the solution space because energy use will continue to increase. Low target values for service levels per capita will be difficult to reach, and more fossil fuels will be consumed, which will result in less available fuel for the transition period. Thus, the challenge for organizational change is set from the viewpoint of natural resources.

6.6. Ways to Forward the Backcasting Approach

6.6.1. Vision and Perception for Desired Futures

The backcasting approach, which is highly solution-driven, is expected to play an important role in specifying appropriate measures, strategies and roadmaps from technological and policy viewpoints. This approach is concerned with shared goals that lead to sustainability using available advanced technologies and is coupled with increasing numbers of future visions with specific targets and a variety of measures derived from backcasting scenarios, which could enable us to make the transition to sustainability (Figure 6.4). However, difficulties will still occur as society makes transitions that will enable us to achieve our mid and long-term goals. Based on the study of Mulder and Biesiot (1998), backcasting develops future visions and fosters power and learning among concerned parties by implementing the pathways. Backcasting will enhance the associated integrated image of probabilities and bottlenecks to scale back emissions and supply insights regarding the operation of and opportunities for establishments and organizations.

Interactive learning is vital because of the state of affairs tools. First, it is important to expend vital thoughts and resources on planning. Consequently, the management of the processes through that user participate in the project. Second, most interesting types of social engagement and learning go through particular backcasting processes that may function as inputs for social interactions and learning that may extend to the far sides of the situations themselves (Robinson, 2003). Dreborg (1996) thought of backcasting primarily as a method for reinforcing creativity regarding long-run developments and policy. To induce property, a general modification might be higher than a progressive improvement in sure cases. Backcasting could support the policy method by exploring important selections and investments that must be created. Third, backcasting may result in the interactive development of a long strategy, and applying the strategy will offer an organization or system with a competitive advantage. Moreover, the capabilities of the backcasting approach for mixed analytical and democratic methods with the roles of computer models serve as mediators between scientific and civic cultures. In addition, backcasting is straightforward and fun for participants, which can result in a new approach, participatory backcasting. Participatory backcasting is an approach with the potential to explore and value possible system innovations towards property.

Thus, participatory backcasting could be a nice facilitator for exploring new solutions to advanced sustainability issues and may be observed as a social experiment. Participatory backcasting manages to determine a learning impact that is significant at the technical and cognitive levels of study (Van de Kerkhof, 2004). However, the implementation of backcasting has several weaknesses because the function of the long run image is not continuously clear. In most definitions of the method, the creation of one or additional fascinating future images accounts for the necessity of backcasting. Although we do not denying this finding, the precise worth and performance of future pictures for backcasting are intrinsically problematic. Although most literature is focused on the development of fascinating future pictures that require being in and of itself joined to a following one, backward analysis uses feasibility evaluation of the desired trajectories towards reaching the desired future. If not, the tool would potentially seem ideal.

6.6.2. Participatory Attitudes

Quist and Vergragt (2006) developed a methodological framework for participatory backcasting by proposing the following five stages: strategic problem orientation; construction of sustainable future visions or scenarios; backcasting; elaboration, analysis and definition of follow-up and (action) agenda; and embedding of the results and generation of follow-up and implementation. These five stages consist of design, participatory, and analytical tools and management, coordination, and communication tools. The backcasting approach is not only multidisciplinary but also transdisciplinary. Additionally, different types of goals can be distinguished regarding the future vision and its requirements, the stakeholder process and the analytical results.

6.6.3. Understanding the Transition

Mayer (1997) explained that a wide range of methods and tools are necessary in a participatory backcasting framework and that four groups can be distinguished to form the outline of a toolkit. The first group includes participatory tools and methods and consists of all tools and methods that are helpful, involve stakeholders and generate and guide interactions among stakeholders. Common tools include workshop tools for generating stakeholder creativity, tools for helping stakeholders in specific backcasting activities and tools for participatory vision and scenario construction. The second group in the toolkit consists of designing the tools and methods for scenario construction and to provide details regarding the systems and processes of the design tools. Third, backcasting involves analytical tools and methods that are not only related with assessing scenarios and designs but also address the process analysis and evaluation, stakeholder identification and stakeholder analysis methods. Fourth, backcasting requires management, coordination and communication tools, including methods for communication, the management process, and shaping and maintaining stakeholder networks that originate from the backcasting study. Each stage of the backcasting approach generally requires tools and methods from all four categories, and different tools and methods are likely involved during different stages.

The transition to a sustainable society requires the integration of multiple levels of innovations that not only lead to drastic changes in technology but also to overall changes in industry and society. Arguments regarding appropriate implementation of advanced technologies in society and their connection with social targets, which make the transition to sustainability possible, have been lacking. Particularly, the research framework and systems in place to help us bridge existing research seeds (or technologies) with shared visions at multiple levels are still weak.

The literature on backcasting demonstrates a considerable goal orientation. However, focus is placed on goals connected with desirable future states. Backcasting is observed as a participatory and process-oriented approach conducted over a limited time. Therefore, goals should not only account for desirable futures, but also processes. Then, goals in backcasting studies can include the generation of normative options for the future and place these options on agendas in relevant arenas, such as future visions or normative scenarios, follow-up agendas that contain activities for different groups of stakeholders and contribute to bringing about the desired future, and stakeholder learning regarding the options, consequences and opinions of others.

6.6.4. Trajectories for Desired Futures

Furthermore, future work regarding the backcasting approach should focus on the following three aspects: (1) improve the linkage between the development of future images and the feasibility analysis of these future images in backcasting, (2) further clarify and improve the roles of scientific models and information to encourage the participants in the backcasting exercise to explore environmental, economic and social aspects of the future, and (3) explore opportunities to improve the quality of the argumentative process in backcasting by using backcasting in combination with other methods.

Future studies that address these challenges to stimulate effective and smooth cooperation with interdisciplinary collaborations among researchers would be useful, particularly for those working on individual technologies at the micro-level (e.g., the engineer) and or involved in the vision aspect (e.g., the economist). It is important to nurture and incentivize researchers who could play important roles in meso-level research with broad perspectives. The other implementation should include in-depth knowledge of technologies and their applications in society. Furthermore, it is important to strengthen the cooperation among various stakeholders, including universities, industries, government and citizens, to successfully conduct meso-level research. Although participatory backcasting can be characterized as a complex and complicated approach, it is possible to teach backcasting in higher education.

6.6.5. Modeling and Tools

Backcasting as an alternative approach for understanding sustainability and provides more vision than prediction. Backcasting has wider implications for environmental and social responsibilities by using a more circumspect appreciation of current knowledge and more flexible and responsive applications. When performing backcasting, the common tools used for modeling include fuzzy theory, genetic algorithms, and neural networks. An example of such an integrated spatial simulation toolbox that combines dynamic and spatial computations is the Spatial Analysis and Modeling Tool (SAMT) (Wieland et al., 2006). The SAMT is an interactive spatial simulation system that serves as a framework for including different models. In SAMT, a link to a neural networks module extracts models from data, and a link to a module for fuzzy modeling allows fuzzy cognitive mapping without expert knowledge. Backcasting offers a useful method for validating a model for long-term ecological and environmental research. In the following sections, the basic theories and tools used for the system approaches are discussed.

System approaches are useful methods for analyzing and finding solutions to nonmathematical models to attain sustainability. Computer performance has increased dramatically in the past 30 years, and soft computing, which is the computation theory supported by computers, has advanced. Soft computing is frequently used in the systems approach. Thus, we introduce some theories of soft computing with examples.

6.6.5.1. Fuzzy Theory

The basic principle of fuzzy theory is constructed by using the fuzzy set advocated by Zadeh (1965). In research of human thinking and the modeling process, Zadeh built a theoretical system by using accurate mathematical methods to describe fuzzy phenomena. A fuzzy set can address a medium grade between existence as "1" and nothingness as "0" for a conventional set. By using the membership function of fuzzy theory, human feeling can easily be applied to the control system and the expert system. Fuzzy theory contributes many practical examples in a variety of engineering fields. In the fuzzy theory, the most important aspect is the "membership function". The difference between a fuzzy set and a classic set is that a fuzzy set uses the membership function and the degree of an element that belongs to the set can be calculated.

For example, medium height (from 165 cm to 175 cm, as an example) in a crisp set is categorized as 1 or 0 on the graph. However, the 164 cm heights are categorized to 0 in the crisp set (Figure 6.5). The result is not familiar because little difference exists between heights of 164 cm and 165 cm (the same occurs between the heights of 175 cm and 176 cm). If we use a grade between 1 and 0, then our categorization of height by the language (or word or expression) can be increased more precisely than by using a crisp set. The membership function of a fuzzy set is an applicable method for expressing human feeling by applying continuous grades of between 0 and 1, as shown in Figure 6.5 (right). The grade *i* for each height *n* of the membership function indicates the ratio of reversion to a fuzzy set. Basic arithmetical operations of the membership functions are affected.

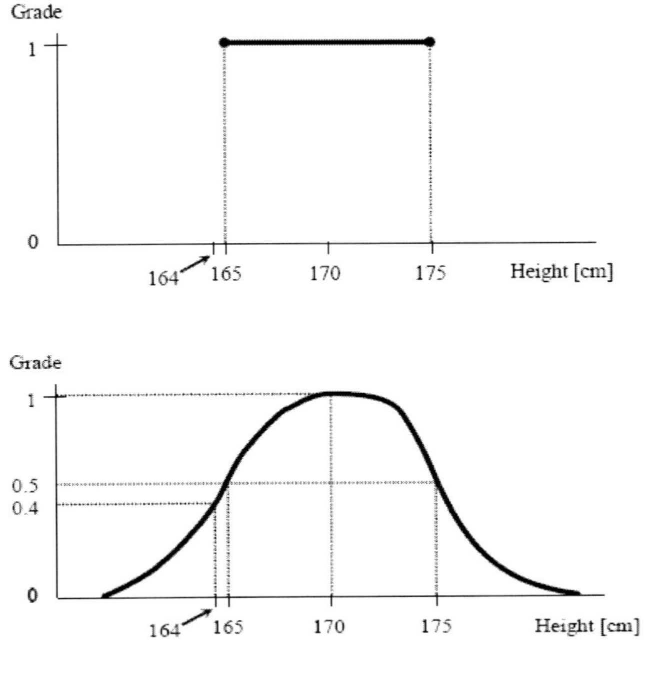

Figure 6.5. Membership function of middle height in the Crisp Set (left) and Fuzzy Set (right).

6.6.5.1.1. Language Variable Application

Language variable application is a unique concept of fuzzy theory. For example, when considering "age", the value of the language variable application is constructed with basic words, such as "young" or "old" and modifiers such as "very" or "more or less" (Figure 6.6). Complicated expressions of the language variable application can be expressed using combinations of the language hedge, such as "very+ very+ young". The language variable application is not an objective index. Users of fuzzy theory apply the membership function of the language variable as a subjective index.

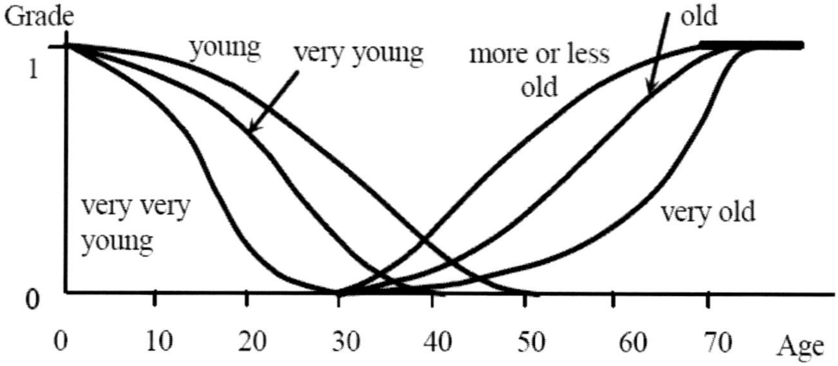

Figure 6.6. Example of the membership function of the language variable applied to human age.

6.6.5.1.2. Inference

Fuzzy inference is an important aspect of fuzzy application. Fuzzy sets A and B are used according to the if-then rule as follows: if X is A and Y is B, then fuzzy inference deduces that B' is an output of Y and B' is an input of A' based on the approximate inference of the control system with fuzzy expressions of human feeling. The first part of the if-then rule is fact, and the second part is the result. The inference process is divided into the three following stages: 1) calculate the conformity for the former part of the if-then rule based on the input condition of fact, 2) calculate each output result based on each if-then rule using the conformity, and 3) calculate the final result by integrating all of the output results. As an example of an if-then rule, the fuzzy inference is calculated as follows:

if-then rule if: space between a car and another car is narrow, then: reduce speed ("normally").

input condition of fact: space is "slightly" narrow.

---

output result: reduce speed "slightly ".

For example, when two if-then rules with two former parts and one after part exist, the fuzzy inference is calculated as follows (Figure 6.7):

Rule 1: if x is A1 and y is B1, then z is C1.
Rule 2: if x is A2 and y is B2, then z is C2.
*fuzzy set: A1, B1, A2, B2, C1and C2.

*input: x, y (input condition of fact: x0, y0), and *output: z (output result: z0).

During the first stage of the inference process, conformity of the former part of the if-then rule by the input (x0, y0) is calculated as follows because the term "and" connects the two former parts:

Conformity of the former part of if-then Rule 1: W1 = μA1(x0) AND μB1(y0).
Conformity of the former part of if-then Rule 2: W2 = μA2(x0) AND μB2(y0).

*μA1(x0), μB1(y0): value of grade in the membership function based on input (x0, y0).
*AND: Adoption for small value.

In the second stage, each output result based on each if-then rule using the conformity is calculated by the head cut of the membership functions, μC1(z) and μC2(z), for the following portion of the inference rule. The level of the head cut is decided based on grades of W1 and W2.

output result of Rule 1: μC'1(z) = W1 AND μC1(z).
output result of Rule 2: μC'2(z) = W2 AND μC2(z).

In the third stage, the membership function of the fuzzy set C was expressed as a combination of μC'1(z) and μC'2(z) as follows:

Membership function of fuzzy set C: μC'1(z) OR μC'2(z).
*OR: whole area of the membership function.

The gravity point of μC'1(z) OR μC'2(z) is calculated for one output result of z0 by integrating all of the output results.

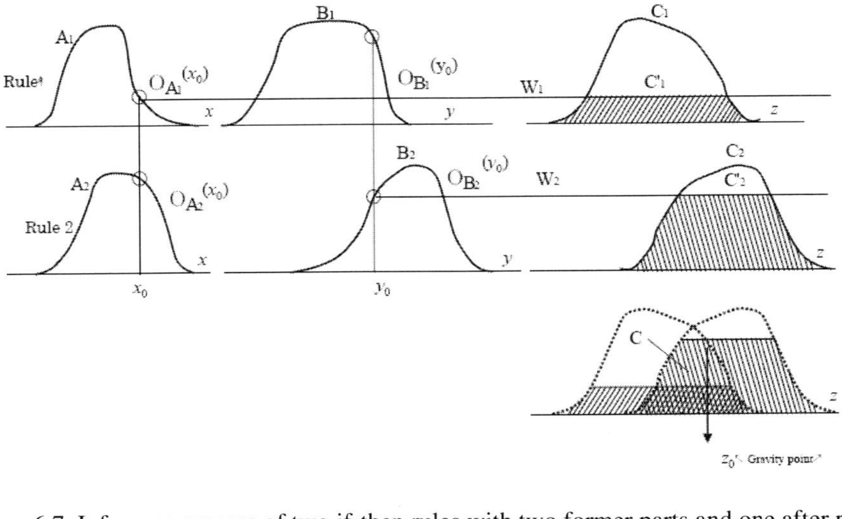

Figure 6.7. Inference process of two if-then rules with two former parts and one after part.

Fuzzy theory can be used to accept the fuzziness of the problem. This theory can be applied to engineering applications by using a theoretical model of human feelings. It is highly possible that a solution can be found for the real world, but computers cannot find all of the solutions for human activities because humans have a brain and use it for decision-making. To overcome this limitation, the neural network theory has been practiced to imitate the neurons in the human brain. The dynamic models and sustainability issues considered the neural network theory for simulations with multiple inputs and single outputs.

6.6.5.2. Neural Network Theory

The urban sustainability indicators that are discussed in Chapter 2 need to progress with dynamic simulation models that consider spatial scales. The backcasting model, where the model output is set to represent a target for sustainable development, is one example. Neural networks are used for simulating expert systems that are embedded within local knowledge and dynamic models. Neural network theory is constructed using computer algorithms by imitating brain cell (neuron) function. Neural network researchers advocate a multi-inputs and single output model based on mathematics as the first stage of neural network theory (Figure 6.8). For example, each neuron has some synapses that connect with other neurons in the body. This information is transferred across the synapse to other cells by communicating with chemicals. The Artificial Neural Network (ANN) is an open concept that allows for the acquisition of new knowledge to provide solutions to solve problems.

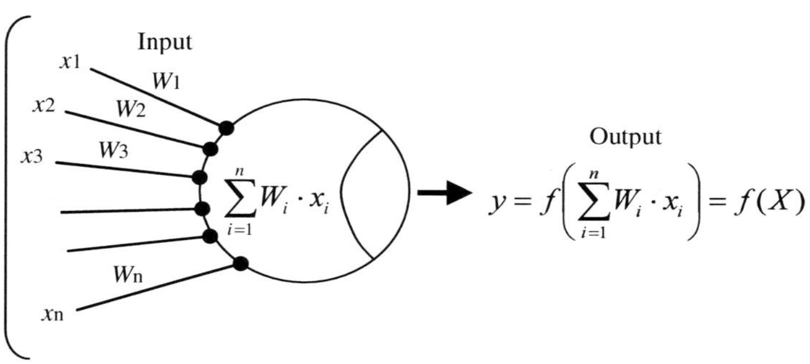

Figure 6.8. Multi-inputs and single outputs of a single unit of the Neural Network model.

The Hopfield and back-propagation models are typical models used in neural network theory. The Hopfield model is applied to the traveling salesman problem of a NP (nondeterministic polynomial), as well as linear programming, Fast Fourier Transfer (FFT), and other applications. The Hopfield model is based on the spin model of physics. A single layer is used for the Hopfield model, and its weights are not changed during the calculation. In contrast, the back-propagation model is primarily applied for pattern recognition, and multiple layers are used in the back-propagation model. The weights of the back-propagation model are changed based on the instructor signal of supervised learning during the calculation. The single unit of the neural network model has multi-inputs and a single output as function of the synapse (Figure 6.8).

When the total value of the input signals multiplied by each weight overflows into the threshold function (as in the left graph and equations of Figure 6.9), "1" of the output signal is gained. The output value of the threshold function is not continuous from "0" to "1". In the program for neural network theory, a sigmoid function is used to guarantee continuous changes instead of the threshold function, which executes the calculations smoothly in the program. The changes in threshold variations are determined from the value of μ_0 in equation (1).

$$y = f(X) = \frac{1}{2}\left\{1 + \tanh\left(\frac{X}{\mu_0}\right)\right\} \tag{6-1}$$

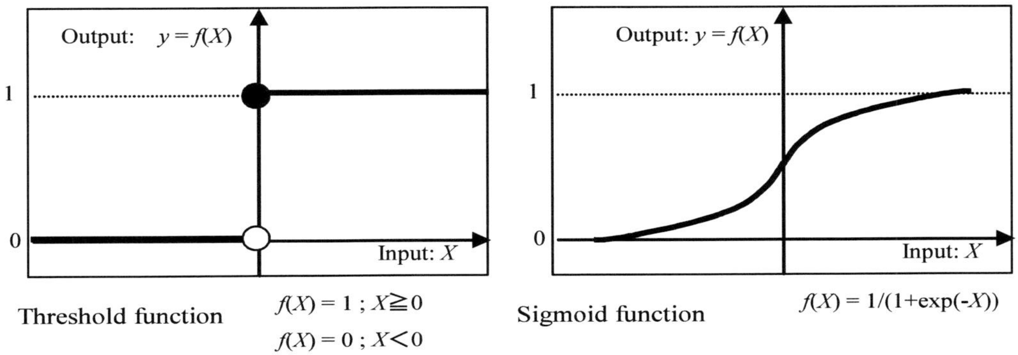

Figure 6.9. Examples of the unit response function: *f*.

6.6.5.2.1. Supervised Learning in Perceptron

The perceptron, which is a single unit combination of a neural network model, learns a variety of patterns through supervised learning (Figure 6.10). The weight of the synapse between neurons changes according to the results of the coincidence for the relationship between the input and output. When the correct output is gained from the input, the weight of the synapse is increased to attain a strong relationship. When a false answer is gained, the weight of the synapse decreases. In the mathematics of neural network theory, supervised learning in perceptron results in the classification of information in patterns. However, a simple perceptron only separates the patterns in N dimensions of space to the hyperplane of N-1 dimensions as a linear separation using equation (2) in mathematics theory. Thus, nonlinear separation is not acceptable for the simple perceptron.

$$\sum_{i=1}^{N} Wi \cdot xi - \theta = 0 \tag{6-2}$$

Recently, the use of artificial neural networks has considerably increased for short and long-term predictions regarding environmental sustainability issues. The Hopfield and back-propagation models are the most common. In the following section, summaries of the neural network Hopfield and back-propagation models are presented.

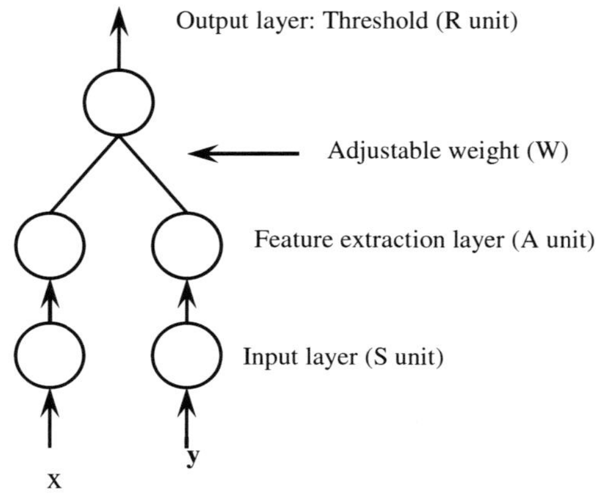

Figure 6.10. Simple perceptron.

6.6.5.2.2. Hopfield Model

Supervised learning in the perceptron is not applied in the Hopfield model. The weights of synapses between neurons do not change from the initial conditions during the calculations. The convergence theory of the Hopfield model for finding solutions is derived from the energy function of physics. The network structure of the Hopfield model is one layer with all neurons connected (Figure 6.11).

Here, $T_{i,j}$ represents the weight of the synapse from neuron (*j*) to neuron (*i*) ($T_{i,j} > 0$: excitement connection, and $T_{i,j} < 0$: inhibition connection); V_i is the output of neuron (*i*) (0 or 1), I_i is the threshold ($I_i < 0$), and *t* is a discrete time. The output of V_i (t+1) is expressed by setting the application model as follows:

$$x_i(t) = \sum_{j=1, i \neq j}^{N} T_{i,j} V_j(t) + I_i \qquad (6\text{-}3)$$

$$V_i(t+1) = \frac{1}{2}\left\{1 + \tanh\left(\frac{x(t)}{\mu_0}\right)\right\} \qquad (6\text{-}4)$$

If the relationship between $T_{j,i}$ and $T_{i,j}$ ($T_{i,j}$:= $T_{i,j}$) is symmetric and the asynchronous changes of each neuron are guaranteed, the energy *E* (Lyapunov function) (Eq. 5) autonomously decreases and T_{ij} and I_i can be found as follows:

$$E = -\frac{1}{2} \sum_{i,j=1,\ j \neq i}^{N} T_{i,j} V_i V_j - \sum_{i=1}^{N} I_i V_i \qquad (6\text{-}5)$$

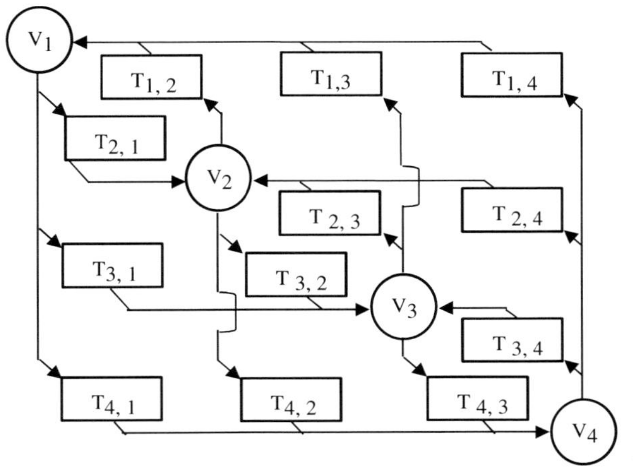

Figure 6.11. Network structure of the Hopfield model (T: weight of synapse and V: cell).

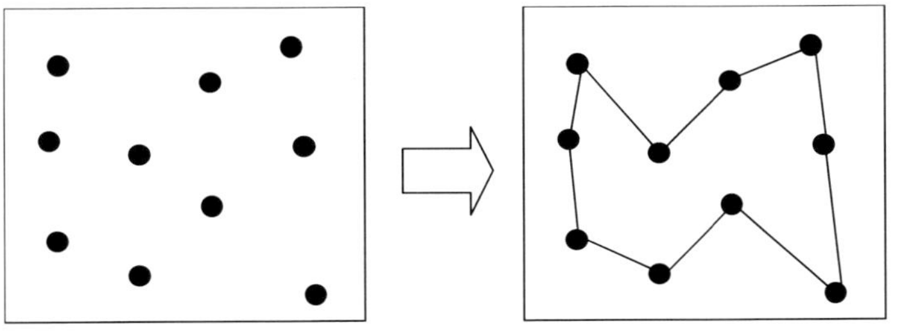

Figure 6.12. Traveling Salesman Problem for 10 cities.

For example, the number of combinations of n people traveling to a city is $n!/2$. If 10 cities exist in this problem, approximately 1.8 million patterns of combination could be considered (i.e., the traveling salesman problem). The Hopfield model can be used to find the minimum number of ways people travel from city to city from many patterns (Figure 6.12).

6.6.5.2.3. Back-Propagation Model

The back-propagation model is a learning algorithm with a multilayer structure. The number of medium layers and the numbers of units for the back-propagation model are determined using experience. In a back-propagation model with two layers, the output O_{kp} of unit k in the O layer is calculated as follows using the connection factor W_{ki} without a threshold and when the pattern, P, is provided as a signal to layer ***I*** (Figure 6.13):

$$O_{kp} = \sum_{i} W_{ki} I_{ip} = W_{k1} I_{iP} + W_{k2} I_{2p} + \cdots + W_{ki} I_{ip} \tag{6-6}$$

The neural network model for minimizing E_t and using the error function, E_p, total error function, E_t, and instructor signal, T_{kp}, is represented as follows:

$$E_p = \frac{1}{2}\sum_{k}\left(T_{kp} - O_{kp}\right)^2 = \frac{\left(T_{1p} - O_{1p}\right)^2}{2} + \frac{\left(T_{2p} - O_{2p}\right)^2}{2} + \cdots + \frac{\left(T_{np} - O_{np}\right)^2}{2} \tag{6-7}$$

$$E_t = \sum_p E_p \tag{6-8}$$

The difference between the error function, E_p, and the connection factor, W_{ki}, is calculated as follows:

$$\frac{\delta E_p}{\delta W_{ki}} = \frac{\delta E_p}{\delta O_{kp}} \cdot \frac{\delta O_{kp}}{\delta W_{ki}} = -2 \cdot \frac{(T_{kp} - O_{kp})}{2} \cdot I_{ip} = -(T_{kp} - O_{kp}) \cdot I_{ip} \tag{6-9}$$

where ΔpW_{ki} is the correction amount of W_{ki} based on the steepest descent method for input pattern P for minimizing E_p and is calculated as follows:

$$\Delta_p W_{ki} = -\alpha_1 \left(\frac{\delta E_p}{\delta W_{ki}} \right) = \alpha_1 (T_{kp} - O_{kp}) \cdot I_{ip} \tag{6-10}$$

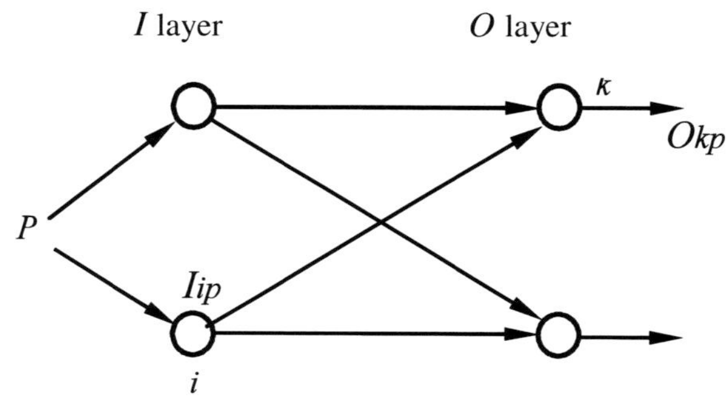

Figure 6.13. Structure of a two-layer model.

In back-propagation models with more than two layers, the integration of the multiplying output, y_i, from unit i (input layer) with the connection factor, w_{ji}, results in the input value of unit j (middle layer). In addition, the integration of the multiplying output, y_j, of unit j (middle layer) with the connection factor, v_{kj}, results in the input value of unit k (output layer) (Figure 6.14). A sigmoidal function f with a threshold is used for this calculation. The total input for a neuron j is calculated using the output of neuron y_i and the weight of the synapse as follows:

$$x_j = \sum_i y_i w_{ji} \tag{6-11}$$

Similarly, the total input for a neuron k is calculated using the output of neuron y_j and the weight of the synapse as follows:

$$x_k = \sum_j y_j v_{kj} \tag{6-12}$$

Many signals in back-propagation models are considered using calculation data from the input layer to the output layer and through the middle layer. The output results of the output layer are compared with the instruction signal. If the output results are correct, the weights of the synapse are strengthened. If the output results are false, the weights of the synapse are weakened. Changing the weights can minimize the differences between the output x_k of layer k and the instructional signal d_k as follows:

$$Minimize \quad E = \frac{1}{2}\sum_k (d_k - x_k)^2 \tag{6-13}$$

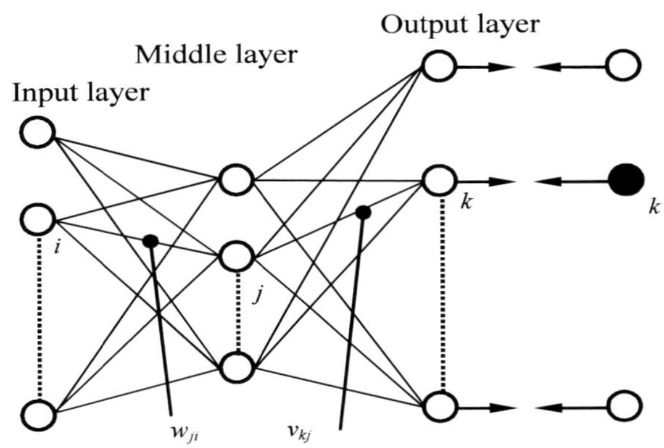

Figure 6.14. Structure of a three-layer model.

The weights of the synapses change from the output layer to the input layer and through the middle layer during the learning process due to supervised learning in the perceptron. Once the neural network is constructed in the model, its weights and biases must be adjusted to produce appropriate outputs. The indicators could influence the learning process.

Neural network theory is a powerful approach for finding solutions for linear and nonlinear models. Humans have acquired useful brain functions through natural selection throughout the long history of life. In the next section, we introduce a genetic algorithm that imitates the evolution of life.

6.6.5.3. Genetic Algorithm

Genetic algorithms use a computing method that imitates the gene model of biology. In the genetic algorithm, a weak gene disappears when the environment is too severe for life. Conversely, a strong gene survives and multiplies through reproduction when the environment is suitable for life. Mutations in genes contribute to evolution. The genetic algorithm is a stochastic simulation method that follows the above scenario, and the genetic algorithm is supposed to be an optimization method of random data that is similar to a Monte Carlo simulation. When using the genetic algorithm for modeling and calculations, the DNA sequence becomes a string of symbols. One life has one individual gene, and a group of life is built of lives. Later, environmental and ecological models for long-term planning adopt this model. For example, this approach was used for the development of multi-objective land planning in the Netherlands (Stewart et al., 2004). This model integrated the genetic algorithm into a complete land use plan as part of a decision support system. In the following sections, the core principles of the genetic algorithm and the requirements for designing the model are discussed.

Modeling (Coding): Information on gene characteristics (DNA sequence) is replaced by a string of symbols in the program by a designer or programmer.

Germination: Some lives in the group are germinated to survive during the evolution process considering the existence of variability between genes.

Evaluation: If the environment is suitable for a life, the life can survive. For the individual, the evaluation function is constructed by coding a string of symbols, which is used after modeling to evaluate the degree of fitness of the environment.

Selection: Some lives with a gene for the next step are selected based on the constant ratio of survivors using the evaluation function.

Reproduction: The total number of lives is reduced by selection, and the newly selected lives reproduce. The reproduction of life only produces life with the same performance as the previous life. Therefore, crossover and mutation are introduced to produce superior genes and increase the total evaluation value of the group. The total number of lives decreases due to selection, and the newly selected lives reproduce. Reproduction of life only produces life that performs the same as the previous life. Next, crossover and mutation are introduced to produce superior life with the gene and to increase the total evaluation value of the group (Figure 6.15).

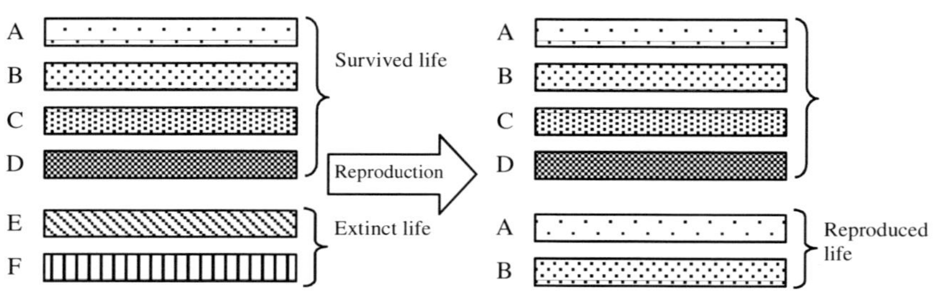

Figure 6.15. Selection and reproduction of life.

Crossover: Two genes between two different lives exchange information in a crossover operation in specific portions of the genes. This operation contributes to produce high performance life that survives the next selection (Figure 6.16).

Mutation: Mutation, often by chemicals or radiation, changes a specific portion in a DNA sequence. In the genetic algorithm at a certain level of probability, a random base changes a specific portion of the string of symbols (gene imitation) (Figure 6.17). Mutation helps produce life with a high performance.

Many applications of the optimization problem exist for nonlinear functions. For example, to find the solution of a minimum value in a nonlinear function, y = f (x), by using a genetic algorithm, 256 steps were set for the x and y axes.

Modeling (Coding): The value of x is an integer from 0 to 255 and is expressed in 8-bit binary. For example, in the string of symbols, the figures "00000000", "11111111", and "00111011" are equivalent to "0", "255", and "59", respectively, for x in the binary system based on binary digits.

Gene setting: Next, 10 genes of life were set for the calculation to avoid the local minimum of nonlinear calculations. Each gene is reflected in 8-bit binary and is initialized using random settings at the beginning of the calculation.

Evaluation: Each gene was evaluated for environmental adaptation. The values of y were calculated using a nonlinear function ($y=f(x)$) after obtaining the value of x for each gene and expressed in 8-bit binary. The smaller value of y is shown in the calculation, and a higher evaluation value is indicated.

Gene Selection: Genes with of the lowest 20% of scores according to the evaluation were eliminated to attain a higher evaluation.

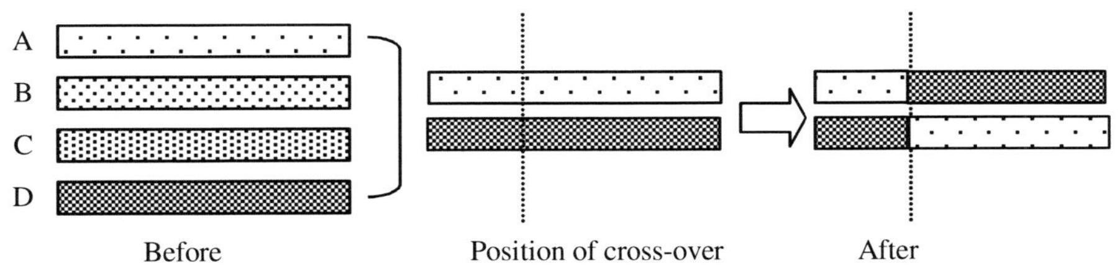

Figure 6.16. Exchange of gene information between two different lives by crossover.

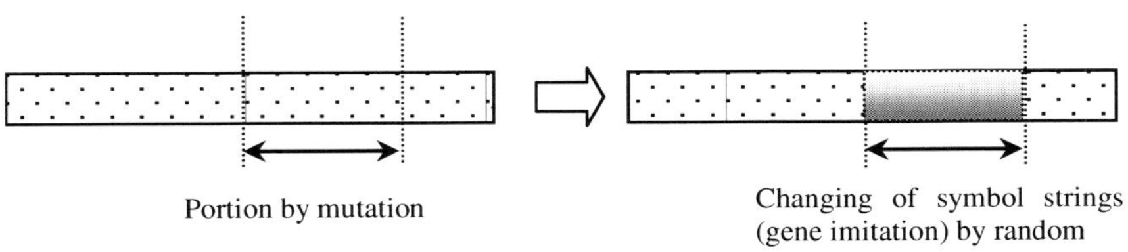

Figure 6.17. Changes in specific portions of symbol strings (gene imitation) by mutation.

Reproduction: After selecting different genes, the groups scoring in the highest 20% of genes were reproduce to compensate for the eliminated genes.

Crossover: In the crossover operation, 40% of one gene and 40% of another gene are exchanged. The position of the crossover section is determined by random operation. For mutation, the lowest 20% of genes were exchanged with new genes based on a probability of 20%.

The left screen shows that the program can be used to find optimum values during limited generations (Figure 6.18). However, sometimes the program cannot find a solution because it is trapped at a local minimum and cannot escape before the end of the generation step. Similarly, sustainable land use planning uses boundaries based on a fast genetic algorithm (Cao et al., 2012). The model was developed to find optimal solutions for land use allocation problems with multiple objectives and constraints. The scenario was developed with trade-offs among economic benefits, environmental, geological and ecological suitability, and urban systems using a genetic algorithm. A genetic algorithm can be applied to many nonlinear models because the computational theory is simple. We can find numerous cases of applying the genetic algorithm to the software and controller. If we can find an excellent new ability in the new life, we may imitate it and construct a new soft-computing theory.

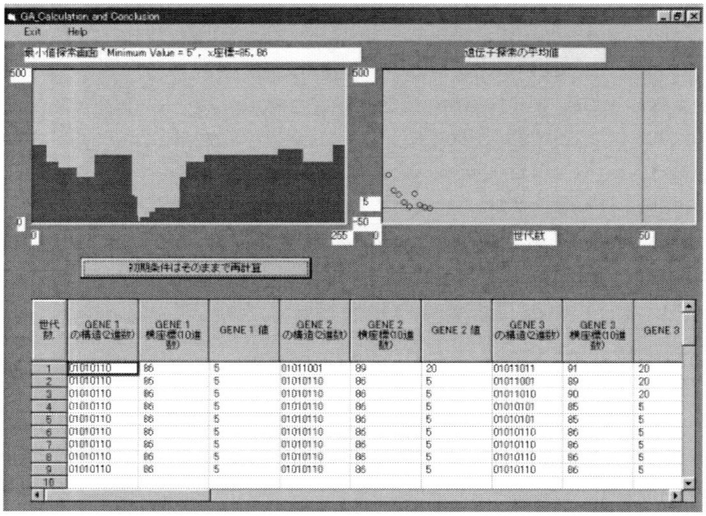

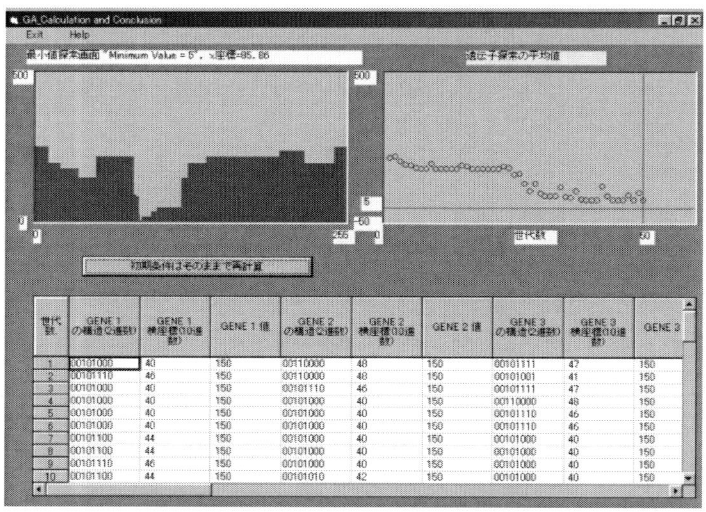

Figure 6.18. Searching for the minimum value of a nonlinear function f=(x) by using a genetic algorithm (left: the program has found the optimum value during limited generation steps; and right: the program has not found an optimum value).

In this chapter, a variety of methods for systemic problem solving are introduced, including backcasting approaches and methods for forwarding the backcasting to understand the future needs, vision and perception of the desired futures. A number of examples and tools commonly practiced by researchers and policy planners are discussed and presented. The systemic approaches are sued to ensure the sustainability of the projects, institutions and programs over the period. Changes in perception involve changes in thinking, which can be thought of as a paradigm shift that is to be introduced. The following chapter introduces system design approaches for renewable energy to find the best scenarios for practicing renewable energy.

Chapter 7

System Design for Utilization of Renewable Energy: System Informatics and Decision Support Systems for Achieving Sustainability in Renewable Energy

The systems approach is a holistic approach for analyzing feedback relationships between the parts of a system and the whole system and for harmonizing the relationships. This method is useful for predicting the bio-resources and renewable energy that is required for achieving sustainability. Several important contexts of system dynamics and backcasting are described relative to bioenergy and system hierarchy for selecting the best practice methods described in the previous chapter. Particularly, bioenergy is closely related to agriculture, forestry, the environment, and society. Thus, this analysis must account for not only bioenergy but also human thinking and human relationships to maintain sustainable bioenergy production while considering food production and environmental protection. This bioenergy analysis can yield a variety of results, including optimism or pessimism towards introducing bioenergy policy and a recommendation to refrain from introducing a bioenergy policy. The variety of evaluation results for bioenergy utilization exist because the boundary conditions of the bioenergy production system are not clear. Thus, we are unable to judge the evaluated bioenergy efficiency based on its consequences. This chapter proposes a method for measuring energy use that considers food production based on system and thermal engineering. This method uses a Life Cycle Assessment (LCA), an inventory analysis of agriculture, linear programming for optimization, system dynamics for renewable energy production, and a system design for exergy and energy. The simulation parameter and procedures are presented to increase the number of skilled hand-on-exercises to understand the system dynamics.

7.1. Inventory Analysis of LCA

7.1.1. Inventory Analysis

The LCA can be used to measure the environmental impacts of a product in a particular setting within the purpose and boundary conditions of the LCA. The LCA inventory analysis measures energy consumption and other emissions to determine the environmental impacts of each stage, including supply, drafting, planning, manufacturing (mining and import), recycling and disposal. Typical methods of inventory analysis include the "interindustryrelations table method" for macro analysis and the "build-up method" based on the accumulation of environmentally hazardous emissions in each component of the process. The direct energy used in food and biomass production can easily be measured and calculated as heat because the amounts of gasoline, diesel, kerosene, and electricity are usually recorded precisely at university farms and research institutes. However, if detailed data regarding energy consumption in the manufacturing process are obtained, the build-up method is effective for calculating the indirect energy inputs through materials, such as agricultural machinery and facility construction materials. For example, Lee et al., (2000) studied the inventory analysis of agricultural machinery in the manufacturing process using the build-up method. The inventory analysis is based on determining energy consumption, with CO_2 emissions serving as a popular component of energy consumption for measurements in LCA research regarding food and bioenergy production. Pimentel (1996), CIGR (1999), and other researchers and organizations have conducted inventory analyses for many different energy crops. Furthermore, several Japanese researchers have studied mechanized rice production. However, very few studies have analyzed the EPR (energy profit ratio) while considering indirect energy in the manufacturing processes of machinery and materials for agriculture due to a lack of clarity regarding the amount of energy consumed in the manufacturing process.

7.1.2. Boundary Conditions

In general, LCA requires boundary conditions. The boundary concept of LCA is "from the cradle to the grave" of a product. However, it is nearly impossible to gather all data for a product "from the cradle to the grave." Thus, the boundary area for inventory analysis must be narrowed to a range that is feasible for data collection. The inventory analysis method for mechanized production is proposed to account for energy consumption in the process of manufacturing agricultural machinery and materials (Noguchi and Saito, 2008). Energy consumption and CO_2 emissions were calculated at the Utsunomiya University Farm (Located in Tochigi prefecture, Japan) by using an inventory analysis. The EPR for rice paddy production was compared between previous and calculated data.

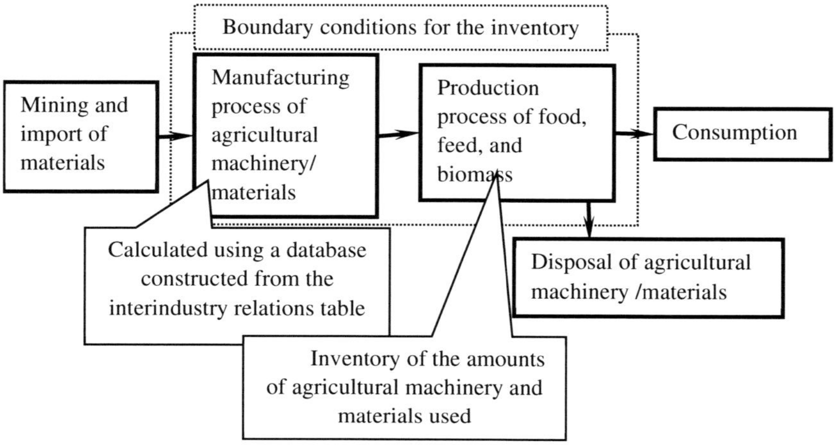

Figure 7.1. Boundary conditions of the inventory analysis for mechanized rice production.

However, it is difficult to obtain detailed information regarding the manufacturing processes of all machinery and materials when using the build-up method because detailed, indirect energy data are rarely published. The inter industry-relations table method was judged to be an appropriate and useful method for calculating the indirect energy of inventory analysis. In addition, the energy consumed by facilities and infrastructure for rice production was removed from the inventory analysis in this example. Thus, the "manufacturing process of agricultural machinery and materials" and the "production processes of food, feed, and biomass" were used as boundary conditions for the inventory analysis (Figure 7.1).

7.1.3. Data Arrangement

A useful database of data obtained from inter industry-relations tables is being developed to enable the application of an inventory analysis (JEMAI 1999, Inaba, 2005). Inventory analysis can be performed within the range of the statistical variability of the data in the database after obtaining the indexes, such as the weights or prices of the materials and machinery used for rice production. The total amounts of energy consumption and emissions during the manufacturing process can be calculated by the database. All secondary energy required to move the machinery, such as distilled fuel or electricity, was added to the indirect energy category of the inventory analysis.

The supply of raw materials, the consumption of food and energy, and the disposal of machinery and materials are outside of the boundary conditions for this inventory analysis. The total amounts of energy consumption and CO_2, SOx, and NOx emissions were calculated by summing the inputs and multiplying them by the corresponding conversion coefficients (representative value) in the database. The conversion coefficients (representative values) in the database (Table 7.1) were used to calculate the energy consumption and the CO_2, SOx, and NOx emissions that occurred during the manufacturing process. The amounts of each item used in the production process were determined from a diary log after categorizing the items as fertilizers, germicides, insecticides, germicides and insecticides, herbicides, fuel production, machinery manufacturing, or fuel use (Table 7.2).

Table 7.1. Conversion coefficient (representative value) in the database (energy consumption and emissions of CO_2, SO_X, and NO_X for the production process of machinery, materials, and fuel)

Item	Unit	Energy (MJ)	CO_2 (kg)	SOx (kg×10^{-3})	NOx (kg×10^{-3})
Compost (domestic animals)*	\10,000	562	45	169	43
Phosphorus fertilizer	t	8,156	868	3,214	954
Mixed fertilizer / standard	t	8,594	934	3,684	898

Table 7.1. (Continued)

Item	Unit	Energy (MJ)	CO_2 (kg)	SOx (kg×10-3)	NOx (kg×10-3)
Mixed fertilizer / NK	t	8,880	965	3,806	927
Fertilizer (others)	1,000 t	535,058	41,048	135,085	40,868
Germicide	t	72,877	7,903	21,871	7,153
Pesticide (agri. chemical)	t	64,161	6,958	19,255	6,297
Germicide & insecticide	T	36,016	3,906	10,809	3,535
Herbicide	t	70,539	7,650	21,169	6,923
Diesel (production)	kL	9,638	1,040	1,753	843
Kerosene (production)	kL	6,756	729	1,228	590
Gasoline (production)	kL	14,590	1,575	2,653	1,276
General machine	\10,000	828	65	329	77
Agricultural dryer	Unit	38,878	3,062	15,457	3,615
Track	\10,000	710	58	345	61
Trailer with engine	Unit	61,696	4,822	26,780	5,748
Diesel (use)	L	38.6	2.6444	18.3	6.34
Gasoline (use)	L	35.2	2.3587	8.2	3
Kerosene (use)	L	37.3	2.5284	0.09	—
Electricity (production)	KWh	9.44	0.547	0.29	0.22

*Compost (domestic animals), 3,000/t (National Domestic Animal Lab. Japan)

Table 7.2. Input materials for rice production

Item	Material and energy	Amount	Database category
Seed	Seed	184.5 kg	—
Fertilizer	Fertilizer	92,000.0 kg	Fertilizer (domestic animals)
	Phosphorus	1,400.0 kg	Phosphorus fertilizer
	Chemical 10-18-16	2,325.4 kg	Mixed/Standard
	Chemical 5-20-20	100.0 kg	
	Chemical for nursery 5-5-5	3.2 kg	
	NK chemical 18-10-16	470.0 kg	Mixed/NK chemical
	Paste fertilizer 11-13-11	666.0 kg	Fertilizer (others)
Germicide	Type A	1,650.0 mL	Germicide
	Type B	800.0 mL	
Germicide	Type C	1,460.0 mL	Germicide (agriculture)
	Type D	49.0 kg	
Germicide & insecticide	Type E	8.0 kg	Germicide & insecticide

Item	Material and energy	Amount	Database category
Herbicide	Type F	14.0 kg	Herbicide
	Type G	10,000.0 mL	
	Type H	10,000.0 mL	
	Type I	4,000.0 mL	
	Type J	4.0 kg	
	Type K	12.0 kg	
Fuel	Diesel	599.4 L	Diesel (production, use)
	Gasoline	123.5 L	Gasoline (production, use)
	Kerosene	450.0 L	Kerosene (production, use)
Electricity	Pump for ground water lifting (4 kW), total working time: 225 h	900.0 kWh	Electricity (production)
	Pump for water recycling (3 kW), total working time: 225 h	675.0 kWh	Electricity (production)
	Grain dryer (2.38 kW +2.71 kW), total working time: 28.1 h	143.0 kWh	Electricity (production)

Source: (Utsunomiya University Farm, 2001, 7.2 ha).

7.1.4. Analysis for Agricultural Machinery

The purchase price and usage time for the agricultural machinery was obtained from a diary log of farming management and from machinery specifications. Table 7.2 provides an example of purchase price and usage time data for agricultural machinery that was recorded using a diary log at the Utsunomiya University farm in Japan along with machinery specifications. For example, the number of input items in a field, n, the amount of input items, I, to a field, I_i, and the conversion efficiency of the database, D_{ij} (j = 1: energy consumption; j = 2: CO_2 emissions; j = 3: SO_X emissions, j = 4: NO_X emissions), were used to calculate the total amount, T_j.

$$T_j = \sum_{i=1}^{n} I_i \times D_{ij} \tag{7-1}$$

Maintenance and repair were included in the inventory analysis when the usage time of the agricultural machinery exceeded its lifetime. Then, the energy consumption ratio of the manufacturing process for new machinery and the maintenance and repair costs in the extended process were the same as the ratio of life time to extended time (Figure 7.2).

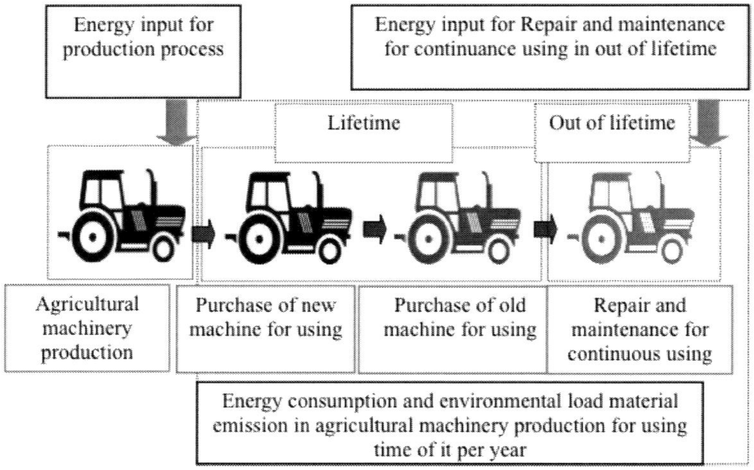

Figure 7.2. Methods for calculating energy consumption and substance emissions for environmental damage in agricultural machinery production and use.

The amount of energy consumed, I_i, and the emissions of CO_2, SO_x, and NO_x were expressed using the units of the database, such as price or weight, M_i, with a life time [h], L_i, and a food or biomass production use time [h], U_i, as shown below.

$$I_i = \frac{U_i}{L_i} M_i \qquad (7\text{-}2)$$

7.2. Linear Programming for Optimization

7.2.1. Linear Programming

Linear programming is a standard method for finding the optimum solution for maximizing or minimizing the objective function under constrained linear equations. For example, a food company produces two types of processed foods, A and B. The main materials used to prepare these foods are fish meat, starch, and vegetable oil. The amounts of these materials in one unit of products A and B are shown (Table 7.3). In addition, this table shows how to determine the number of units of product A or B that should be produced to maximize the total sales of A and B under restrictions in the supply of stock materials and in the unit prices of A and B. The unit prices are 2 USD for A and 3 USD for B, respectively. The number of units is expressed as x_1 for A and x_2 for B. The aim of this analysis is to maximize the total sales for A and B under the constrained equations. Thus, the objective function in linear programming is defined as follows:

$$\text{Maximize: } z = 2x_1 + 3x_2 \qquad (7.3)$$

Constrained equations for the limitations on fish meat, starch, and vegetable oil materials are defined as follows:

$$x_1 + 2x_2 \leq 14 \tag{7-4}$$

$$x_1 + x_2 \leq 8 \tag{7-5}$$

$$3x_1 + x_2 \leq 18 \tag{7-6}$$

Table 7.3. Amount of material required for one product and the unit prices of A and B

	Fish meat	Starch	Vegetable oil	Unit price
A	1 kg	1 kg	3 kg	2 USD
B	2 kg	1 kg	1 kg	3 USD
Stock materials	14 kg	8 kg	18 kg	

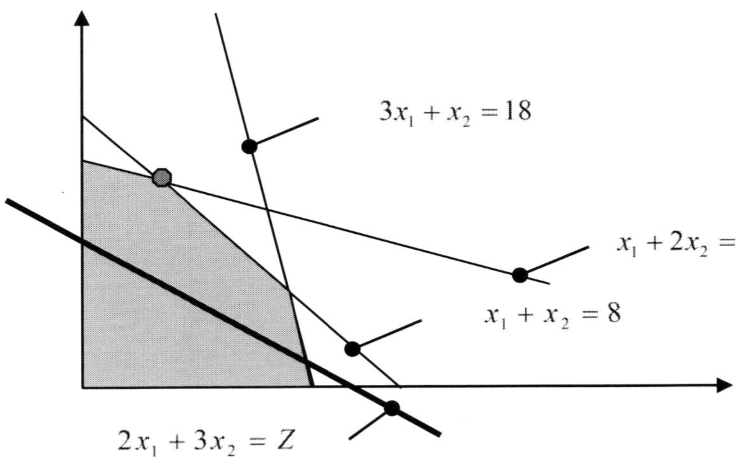

Figure 7.3. Graphical solution for linear programming.

All unit numbers must be greater than 0. Thus, the non-negative condition was defined as follows:

$$x_1 \geq 0, \quad x_2 \geq 0 \tag{7-7}$$

The graphical solution is one method of solving linear programming problems. The graphical solution makes it easy to understand how to obtain a solution in 2 dimensions. When the objective function is solved by the graphical equation (Figure 7.3), the gray color indicates an area that satisfies the constraint conditions. The maximum value of the objective function can be obtained from the parallel movement of the objective function that brings it into contact with the gray area. When the objective function contacts point A, which is one of the peaks of the area where the constraints are satisfied, the value of the x_2 axis section indicates the maximum value. The position of point A is obtained using the following equation.

$$\begin{cases} x_1 + 2x_2 = 14 \\ x_1 + x_2 = 8 \end{cases} \tag{7-8}$$

Finally,

$$x_1 = 2, \quad x_2 = 6 \tag{7-9}$$

The total sales of A and B are expressed as follows:

$$z = 2x_1 + 3x_2 = 2 \times 2 + 3 \times 6 = 22 \tag{7-10}$$

Thus, producing 2 units of A and 6 units of B results in a maximum total sales of 22 USD, which represents the solution for this case. If a linear programming problem involves more than 2 parameters, the graphical solution is not a convenient method and the "simplex method" can be used. Software such as the additional "Solver" function of Excel (Microsoft Cooperation) could be good for solving simple linear programming models.

7.2.2. Simulation for Renewable Energy

Renewable energy is currently receiving a great deal of attention due to the expected decrease in dependence on fossil energy, the creation of new employment, and decreases in global environmental damage. A variety of studies have indicated that biomass energy is one of the most promising sources of renewable energy (Alam et al., 1997; Yamamoto et al., 2000; Mochizuki and Sakoda 2005; Nakata et al., 2005). Noguchi and Misumi (2007) proposed a simulation model for a self-sufficient society and its food supply. In addition, agricultural energy production was used to analyze the competition between the production of food crops and energy crops, which are restricted to certain field utilization scenarios (Figure 7.4).

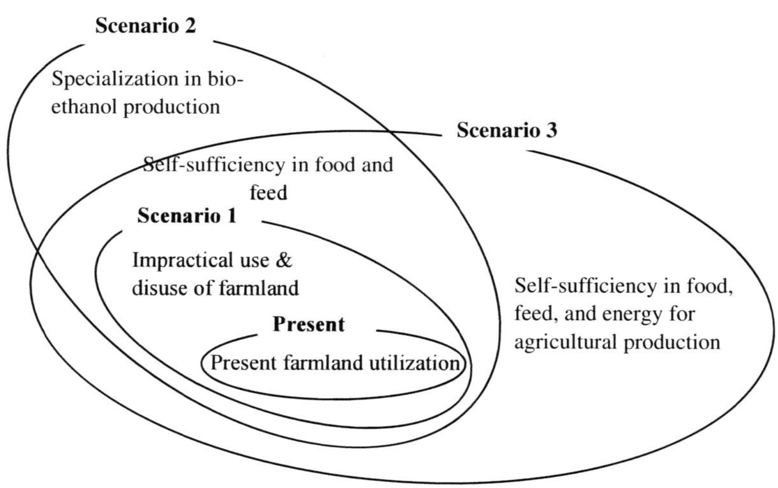

Figure 7.4. Relationships among the scenarios in the simulation model for renewable energy and food production.

Trial calculations were executed using the simulation model under a scenario in which the farmland was used effectively, with high food crop and energy crop production levels for attaining a self-sufficient society. In the simulation model, heavy oil, gasoline, diesel oil, kerosene, coal, LNG, LPG, electricity, bio-gas, and bio-ethanol were used for public welfare, industry, and transportation. Food crop, feed crop, energy crop, and livestock items were determined by referring to the Tochigi branch office (2006).

7.2.3. Simulation by Linear Programming

The model presented in this section aims to optimize a system based on engineering methods, such as mathematics and computing. The model should enable the simple expression of the target. Next, the model requires a clear boundary condition, purpose of optimization, and constraint condition for each module. When linear programming is applied to the model, the objective and constraint functions should be constructed based on the boundary conditions (Konaka, 1997). In the following section, an example is given based on the agricultural residues of farmlands using an objective and constraint function.

The flows of food, energy, and biomass in the simulation model are determined based on the area of farmland in practical use, impractical use and disuse. The objective and constraint functions of linear programming in the simulation model can be used to derive the optimal solution. When all food demands in the target area are filled by the supply, the constraint function is expressed by the following equation based on the average field area of the food crop l, K_l the cropping pattern, x, the area of utilized farmland, S_x, the demand for food crop l per person, Q_l (t/yr), the population, M, and the area of farmland in use for food crop l with cropping pattern x, $S_{x(l)}$.

$$\sum_{x=1}^{x'}\sum_{l=1}^{l'} K_l S_{x(l)} \geq \sum_{l=1}^{l'} Q_l M \tag{7-11}$$

The amount of the energy crop is expressed using the energy crop, n, and the amount of biomass, $R_{ural(n)}$.

$$B_{Rural(n)} = \sum_{x=1}^{x'}\sum_{n=1}^{n'} K_n S_{x(n)} \tag{7-12}$$

Equation (7.12) can be changed to equation (7.13) based on the conversion efficiency in the energy plant by using the amount of energy, k, demanded per person, Q_k. Constraint must be set in equation (7.13) to satisfy the energy demand.

$$f\left(B_{Rural(n)}\right) = f\left(\sum_{x=1}^{x'}\sum_{n=1}^{n'} K_n S_{x(n)}\right) \geq E_{\mathbf{Rural(k)}} = \sum_{k=1}^{k'} Q_k M \tag{7-13}$$

When all feed demand is filled by the supply, the constraint function is expressed using the following parameters: feed crop, m, TDN (Total Digestible Nutrients) yield of feed crop m, K_m, TDN demand of livestock o, R_o, number of livestock, T_o.

$$\sum_{x=1}^{x'}\sum_{m=1}^{m'} K_m S_{x(m)} \geq \sum_{o=1}^{o'} R_o T_o \tag{7-14}$$

When farmland previously under impractical use and disuse is used to produce food crops and energy crops, one constraint condition is that the additional farmland area, S_x, should not exceed the area of the farmland under impractical use and disuse, S_{AB}. Additionally, the farmland areas of the paddy field, $S_{x(Paddy)}$, upland field, $S_{x(Upland)}$, and grassland, $S_{x(\text{Grassland})}$, should not exceed the total farmland area under impractical use and disuse for each category field. Then, the constraint functions are expressed as follows:

$$\sum_{x=1}^{x'} S_x \leq S_{AB} \tag{7-15}$$

$$\sum_{x=1}^{x'} S_{x(Paddy)} \leq S_{AB\_Paddy} \tag{7-16}$$

$$\sum_{x=1}^{x'} S_{x(Upland)} \leq S_{AB\_Upland} \tag{7-17}$$

$$\sum_{x=1}^{x'} S_{x(Glassland)} \leq S_{AB\_Grassland} \tag{7-18}$$

When farmland under practical use and under impractical use and disuse is used for food crop and energy crop production after reexamining all farmland utilization, the total farmland area, S_x, should not exceed the total area of the farmland under practical use, S_{AA} (ha), and impractical use and disuse S_{AB} (ha). Furthermore, the total farmland areas of the paddy fields, $S_{x(Paddy)}$, the upland fields, $S_{x(Upland)}$, and the grasslands, $S_{x(Grassland)}$, should not exceed the total area of farmland under practical, impractical and disuse for each category of field. Thus, the constraint functions are extended as follows:

$$\sum_{x=1}^{x'} S_x \leq S_{AB} + S_{AB} \tag{7-19}$$

$$\sum_{x=1}^{x'} S_{x(Paddy)} \leq S_{AA\_Paddy} + S_{AB\_Paddy} \tag{7-20}$$

$$\sum_{x=1}^{x'} S_{x(Upland)} \leq S_{AA\_Upland} + S_{AB\_Paddy} \tag{7-21}$$

$$\sum_{x=1}^{x'} S_{x(Glassland)} \leq S_{AA\_Glassland} + S_{AB\_Grassland} \tag{7-22}$$

A scenario of minimized cropping area (objective function) is satisfied with food self-sufficiency and can be set up in the present cropping system by utilizing impractical use and disuse of farmland. The objective function in this scenario is to minimize the total area of farmland, S_x, that is categorized as cropping pattern x so that the objective function is expressed as follows:

$$Minimize \sum_{x=1}^{x'} S_x \tag{7-23}$$

When a scenario is set to maximize the amount of bio-ethanol production (objective function) by using impractical use and disuse and by maintaining the current two cropping systems of rice (*Takanari*) and rye, the objective function is expressed as energy $E_{\mathrm{Rural}(k)}$ (bio-ethanol: k) as follows:

$$Maximize\ E_{Rural(k)} \tag{7-24}$$

7.3. System Dynamics for Renewable Energy

7.3.1. System Dynamics

"System dynamics" was advocated by Jay Forrster (MIT) as a method for grasping complicated systems based on the structural concepts of "Stock" and "Flow"(Figure 7.5).

In addition, system dynamics represent a simulation method that can be used to analyze and understand complicated systems, such as social systems that use engineering control theory. System dynamics cover enterprise decision-making, public management, public policy, biological modeling, energy issues, environmental issues, complicated systems, non-linear systems, and more.

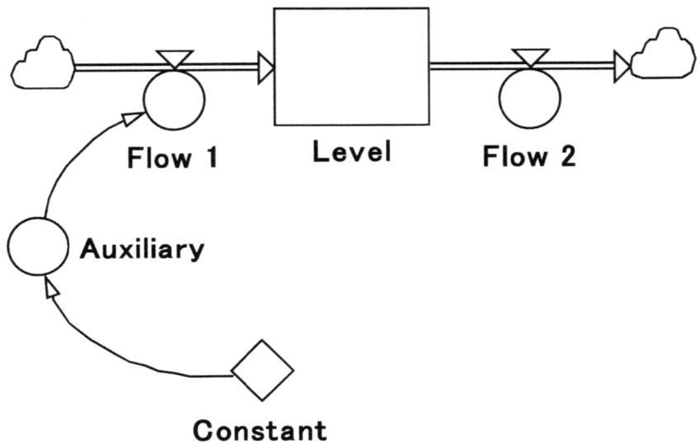

Figure 7.5. Flow, Level, Constant, and Auxiliary in the system dynamics model.

The global dynamics in "The Limits to Growth (1972); A Report for the Club of Rome's Project on the Predicament of Mankind", a research report by the Club of Rome is used as a famous example for the application of system dynamics.

However, this technique has not been applied to biofuels because of ambiguous technology, insufficient economic analysis, biofuel consumption and technological improvements in biofuel production.

7.3.2. Energy Flow Model for System Dynamics

Biofuels are one of the most attractive forms of energy that could serve as a substitute for fossil fuels for automobiles. The transportation sector accounts for 23.7 % of total energy consumption in Japan ($3{,}793\times10^{15}$ J). Almost all of the energy used in the transportation sector is liquid fuel distilled from fossil fuels. Bio-ethanol and bio-diesel have been focused on as the next automobile fuels, and technologies that convert non-edible parts of agricultural crops to second-generation biofuels are under development around the world. However, EVs (electric vehicles) and PHVs (plug-in hybrid vehicles) are already being sold with low fuel consumption and low mileage costs compared with conventional GVs (gasoline vehicles) and DVs (diesel vehicles). In a trial calculation based on the different flows of biomass in energy conversion (Horio, 2007), the total efficiency of EVs is higher than that of GVs or DVs. In this case, electricity is produced by a thermal power gasification plant, ethanol is produced by fermentation and distillation, and each route consumes the same amount of biomass.

An energy flow model with multiple types of biomass and multiple types of energy was proposed for a system dynamics analysis of the possibility of using bioenergy as an alternative energy source in automobiles in rural societies. Six scenarios were investigated, including the utilization of farmlands currently under impractical use and disuse for biomass production and the diffusion of EVs instead of GVs in the Tochigi prefecture. Biomass production, energy plants, and energy consumption were used for the energy flow model to consider recycling energy from biomass production and the energy plant (Figure 7.6). In the energy flow model, the biomass produced from the biomass production module was used for the energy plant. The energy produced from the biomass plant module, $\mathbf{E}^{k}_{RenE}$, was not only used for the energy consumption module in rural areas, $\mathbf{E}^{k}_{RenFC}$, but was also for energy recycling for biomass production, $\mathbf{E}^{k}_{RenAgp}$, and for the energy plant using a time step of k (for example: year). The energy profit ratio was used to consider the energy flow model and to calculate the uniformity of the entire model.

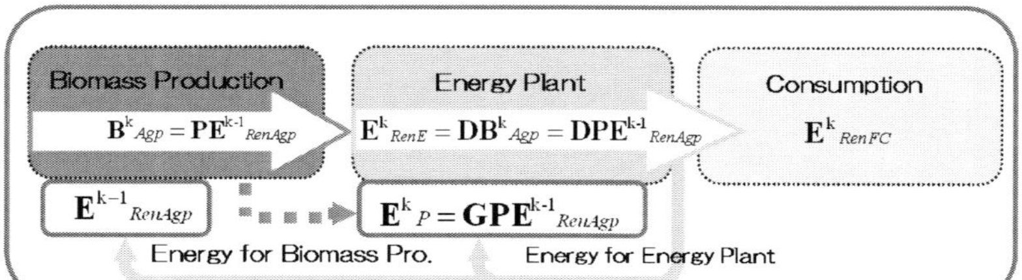

Figure 7.6. Energy flow model considering energy recycling for biomass production and the energy plant.

The conversion coefficient, P, was set to estimate the relationships between the total amount of biomass and the input energy because the amount of biomass production was supposed to be proportional to the amount of input energy for the field. The conversion coefficient, D, was set to estimate the total amount of biomass energy based on the input biomass because the amount of biomass energy should be proportional to the amount of biomass input to the plant. The conversion coefficient, G, was set to estimate the total amount of energy input into the energy plant because it should be proportional to the amount of biomass input into the plant. The surplus energy is used for biomass production and by the energy plant during the next step (next year), which occurs after the produced energy is used for energy consumption in the rural area. Then, the energy flow of recycling is expressed as shown below (the system dynamics model of bioproduction using equation (7-25) and (7-26) is shown in Figure 7.7).

$$\mathbf{E}^{k}{}_{RenE} = \mathbf{DPE}^{k\text{-}1}{}_{RenAgp} \quad (7\text{-}25)$$

$$\mathbf{E}^{k}{}_{RenFC} + \mathbf{E}^{k}{}_{RenAgp} + \mathbf{GPE}^{k}{}_{RenAgp} = \mathbf{E}^{k}{}_{RenE} \quad (7\text{-}26)$$

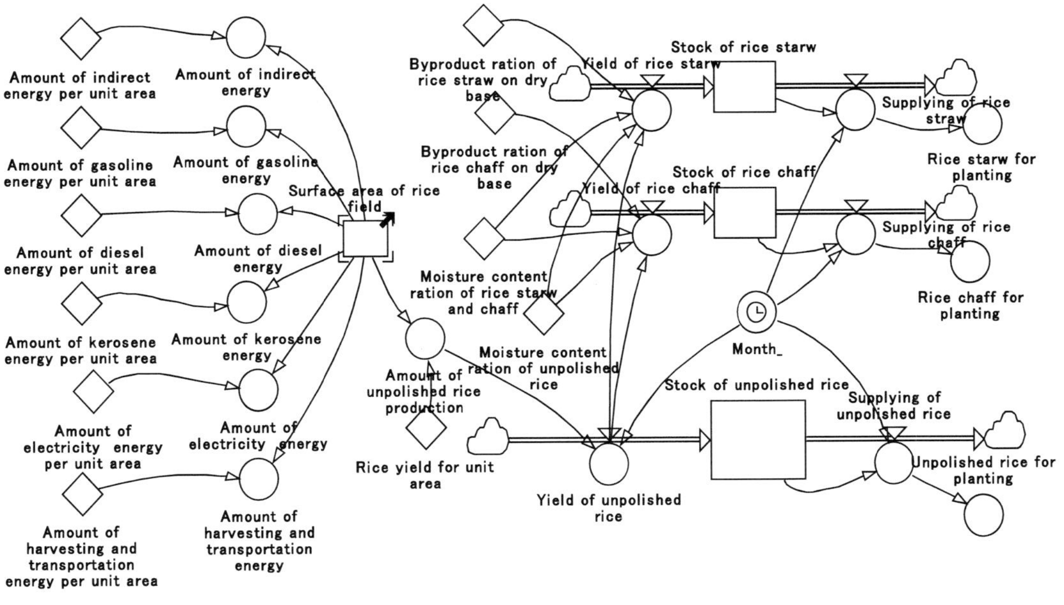

Figure 7.7. System dynamics model of biomass production.

7.4. System Design Approach for Energy Utilization

7.4.1. System Design for Exergy

When a heat source at high temperature and low temperature is not altered by the process of heat transfer, Q [J], the maximum amount of work obtained from this heat is defined by the high temperature (the absolute temperature), T_H [K], low temperature (the absolute temperature), T_L [K], the amount of heat, Q [J], and the Exergy [J] based on the heat efficiency of the Carnot cycle.

$$\frac{E}{Q} = 1 - \frac{T_L}{T_H} \tag{7-27}$$

In another case, the temperature of one resource is nearly stable, and the total amount of heat emitted by a source in response to a small change in temperature, Q [J], is defined based on the mass of the source, m [kg], its specific heat at constant pressure, C_p [J/(kg*K)], the temperature of the resource, T [K], and the surrounding temperature, T_0 [K].

$$Q = mC_P(T - T_0) \tag{7-28}$$

The maximum amount of work acquired from a heat source at temperature T': $Q = mC_PT'$ is defined by considering small changes in temperature based on the Carnot cycle.

$$dQ = mC_PdT'\left(1 - \frac{T_0}{T'}\right) \tag{7-29}$$

A temperature change from T to T_0 results in the maximum amount of work (Exergy) obtained from the system, as defined below. However, 100% of the electricity can theoretically be converted to work, and E/Q is 100%.

$$\begin{aligned} E &= \int_{T_0}^{T} mC_P\left(1 - \frac{T_0}{T'}\right)dT' = mC_P\left[T' - T_0 \ln T'\right]_{T_0}^{T} \\ &= mC_P\left\{(T - T_0) - T_0 \ln \frac{T}{T_0}\right\} \end{aligned} \tag{7-30}$$

7.4.2. System Design for Backcasting

Biofuels have attracted attention as a high-value biomass alternative energy source that could replace fossil fuels and reduce CO_2 emissions. In conventional systems designed for biofuel production, the final benefit is considered after calculating the field area, considering the type of biomass, the characteristics of the biofuel plants, and the characteristics of the biofuel based on a single energy use and a single benefit. From the viewpoint of quality of life, environmental protection, and sustainable development, biofuel production could be introduced by using "backcasting," an approach that identifies policies and programs that connect a defined and desirable future to the present (Oliver and Brooks 2005). In the previous Chapter, backcasting approaches are discussed for a desirable future, especially regarding bioenergy sustainability and final benefit. If final benefit is considered as the goal of the system design for biofuel production, then biofuel type, biofuel plant, biomass type, and field area are ranked based on the applications of multiple types of energy and their final benefits. Next, societal acceptance is considered for analyzing the introduction of the biofuel. The purpose of proposing the evaluation function is to quantify the energy supply and demand, which is defined as the final benefit. The field area for biomass production, the biomass type, and the conversion efficiency of the biofuel plant are used for designing biofuel production systems for total energy efficiency.

To obtain a higher evaluation when the difference between the energy supply and energy demand is nearly zero, the evaluation function, E_{system}[J^2], energy consumption, E_{cons}[J], and energy production, E_{pro}[J], are used for the evaluation function based on the following concept of parameter identification:

$$E_{system} = (E_{cons} - E_{pro})^2 \tag{7-31}$$

Additionally, biofuel production can present over a wide variety of final benefits and energies. Therefore, the evaluation function, E_{system}, can be expressed using the variety of energy, j, energy consumption, E^j_{cons} [J], and energy production, E^j_{pro}[J], as follows:

$$E_{system} = \sum_j (E^j{}_{cons} - E^j{}_{pro})^2 \tag{7-32}$$

After defining the field area of biomass production, S [ha], biomass yield per unit area, Y [t/ha], variety of energy, j, and energy conversion efficiency (including the biofuel plant), η^j [J/t], when E^j_{pro} is proportional to S, the E_{system} can be expressed as follows:

$$E_{system} = \sum_j (E^j{}_{cons} - SY\eta^j)^2 \tag{7-33}$$

Generally, a limited field area of biomass production is cultivated for biofuel production. Minimizing the field area of biomass production satisfied with energy demands for final benefit is one of the best strategies for reducing environmental impacts and maintaining agricultural production. Therefore, equation (7.33) can be differentiated to obtain the most efficient system design for biofuel production by minimizing the field area.

$$\frac{dE_{system}(S)}{dS} = \frac{d\left\{\sum_j (E^j{}_{cons} - SY\eta^j)^2\right\}}{dS} = 0 \tag{7-34}$$

7.4.3. Evaluation using the EPR

The EPR (Energy Profit Ratio), which indicates the ratio of energy output to energy input in biofuel production, is a highly popular standard method for evaluating decision-making. However, energy consumers seek final benefits, such as heating, lighting, and mileage. In addition, it is important to measure the physical quantities of energy for biofuel production to enable an objective determination of the total social benefit. The conventional EPR (Energy Profit Ratio) is calculated using the thermal units [J] of energy production and energy consumption as follows:

$$\mathrm{EPR} = \frac{\sum_j E^j{}_{pro}}{\sum_j E^j{}_{cons}} \tag{7-35}$$

The thermal unit [J] in the EPR is not satisfactory from the viewpoint of available energy because unitti is not expressed as available energy. "Exergy" is defined as the maximum amount of work obtained from the system until the point when the system reaches an equilibrium state. Alternatively, "exergy" is defined as the difference in quality between the different varieties of energy. Thus, $E_{system}{}^{EX}$ is based on "exergy" and was proposed based on exergy theory to evaluation the function E_{system}. Because electricity has 100% exergy efficiency, virtually 100% of electricity can be converted to energy for mechanical work. In contrast, 15°C water in an environment with a temperature of 15°C has 0% exergy efficiency. It is impossible to extract work from 15°C water according to the theory of the Carnot cycle. The ExPR (Exergy Profit Ratio) is the ratio of output energy to input exergy and can be expressed by setting the variety of energy, j, the exergy consumption of j, $E_X{}^j{}_{cons}$ [J], and the exergy production of j, $E_X{}^j{}_{pro}$ [J]. The ExPR can be expressed a follows:

$$\mathrm{ExPR} = \frac{\sum_j Ex^j{}_{pro}}{\sum_j Ex^j{}_{cons}} \tag{7-36}$$

The following section introduces systems-based approaches for using renewable energy and introduces physical properties of exergy to achieve sustainable agricultural development by using renewable energy. Practical, vague and fuzzy subjects should be included in the simulation model, and critical decision-making is required to use the basic concepts of the simulation model. Thus, various researchers have used the decision support model with tolerance power as a method for discussing renewable energy issues.

7.5. Estimation Method of Renewable Energy and Potential

Renewable energy is becoming increasingly important, and energy sources such as solar, wind, hydropower and biomass have become substitute options for fulfilling energy demands in rural sectors. In this rural sector, it is important to employ a renewable energy estimation policy to identify substitute or additional energy sources to decrease dependence on fossil fuel and nuclear energy sources. Geographical distribution can help policy planners increase renewable energy use in rural areas. The smart grid is expected to be the next power network for renewable energy (Ministry of Economy Trade and Industry, Japan 2010a; Mizuho Information and Research Institute, 2010). Thus, we proposed a new estimation method for renewable energy utilization that uses geographical details based on data from geographic and climatic conditions. Furthermore, we identify the shortcomings of the previous study performed by the Regional New Energy Vision (Wakeyama and Ehara, 2009).

A methodology is presented to estimate the potential of renewable energy in rural areas for solar panels, wind turbines, hydropower and biomass. Electric power generation by solar panels (E_S, kWh) is related to the surface area of the solar panel and the accumulation of solar radiation over time and can be expressed as follows:

$$E_S = \int \{H_S(t) \cdot K_{1S}(t) \cdot K_{2S} \cdot \eta_S \cdot A_S\} dt \tag{7-37}$$

where H_s is the solar radiation on the panel area (m^2), K_{1S} is the temperature correction coefficient, K_{2S} is the total correction coefficient, η_S is the conversion efficiency, and A_S is the surface area of the solar panel

Wind power depends on the wind receiving area, air density and kinetic energy. Kinetic energy can be estimated from the mass and velocity of the air. The wind energy potential can be expressed as follows:

$$P_{WT} = \frac{1}{2} \cdot \rho \cdot A_W \cdot V^3 \cdot e_w \cdot 10^{-3} \tag{7-38}$$

where P_{WT} is the total power received from the wind turbines (kW), ρ (kg/m^3) is the air density, A_W is the wind receiving area (m^2), V is the velocity of air (m/s), and e_w is the conversion efficiency. The kinetic energy can be calculated from the following equation:

$$P_{WT} = \frac{1}{2} mV^2 \cdot e_w \cdot 10^{-3} \tag{7-39}$$

where m is the mass of the air. Therefore, the total wind energy potential (E_w, kWh) can be expressed as follows:

$$E_W = \int P_{WT} dt = \int \left\{ \frac{1}{2} \cdot \rho \cdot A_W \cdot V^3(t) \cdot e_w \cdot n \cdot 10^{-3} \right\} dt \tag{7-40}$$

where V is a function of time t, and n is the number of wind turbines. The specific geographical and climate conditions for each of the parameters have been defined in detail. Wind turbines (1,000-kW-rated output) can be installed in selected geographical locations depending on the slope, installation height, and surroundings. The land slope for the turbines should generally not exceed 20°, and installations should not be placed at heights of more than 1,000 m above the ground level. Wind turbines must be installed outside of natural parks and buildings. In addition, turbines should not be installed more than 300 m away from public roads (Wakeyama and Ehara, 2009). Wind turbines are set to capture stable winds by considering the recommended cut-in velocity, the recommended cut-out velocity of the wind, and the recommended distances between the wind turbines. The cut-in and cut-out velocities of the wind and the distance between wind turbines are 4 m/s, 25 m/s, and 10*D*x3*D*, respectively, where *D* is the diameter of the rotor (NEDO, 2008).

Theoretical hydropower (P_H, kW) can be obtained from the potential energy of water (NEDO, 2003). The formula for small hydropower P_H can be expressed as follows:

$$P_H = g \cdot Q_H \cdot H_e \tag{7-41}$$

where g is the gravitational acceleration, Q_H is the amount of water discharge, and H_e is the effective head. The energy potential (E_W) of small hydropower plants can be expressed as follows:

$$E_w = \int \{P_H \cdot \eta \cdot K_W \cdot t\} dt \tag{7-42}$$

where η is the conversion efficiency of small hydropower generation, K_W is the available ratio, and Q_H changes over time t (m^3/s), as demonstrated in the following equation:

$$Q_H(t) = \frac{1}{3.6} \cdot f \cdot r(t) \cdot A_W \tag{7-43}$$

where f is the discharge coefficient, r is the rainfall intensity (mm/h), and A_W is the size of the catchment area ($km^{2)}$. The electric power generation from biomass (E_B, kWh/year) can be determined from the surface area of the field (S_B, ha) used for biomass production (M_B, kg/ha). The energy potential can be expressed as follows:

$$E_B(B) = e_B \cdot P_B(B) \cdot S_B(B) \cdot M_B(B) \tag{7-44}$$

where B is the type of land use for abandoned fields (for energy crop production) and rice fields (for residues, i.e., chaff and straw) and was used to estimate the biomass energy potential.

In this chapter, LCA, linear programming, system dynamics, system design approach, and an estimation method for renewable energy potentials are introduced. The developed method of system design can be integrated with the decision support systems of agricultural production and system informatics to reinforce sustainability planning. To reinforce planning and simulation, backcasting approaches could be implemented to achieve the desired sustainability in agriculture, the environment, and renewable energy. In the following chapter, the system informatics and information communication that consider the sustainability of food and bioenergy are discussed. The informatics indicators are common tools to expand ICT and its sustainability in agriculture and are focused on including sensors and data analytics.

The ICT is one potential method for bridging simulation and reality. The ICT promises to encourage sustainable growth and easy accesses of information throughout the agrarian community and the drive to implement Data to Knowledge. Data to Knowledge resulted from Big Data initiatives as a concept for improving the speed, volume, and variety of data used in agriculture. This Big Data concept is an ICT-based innovation that has driven the economic transformation of precision agriculture from a business perspective. Recently, the growth of ICT has become very fast and its price has dramatically decreased. This growth and price decrease have accelerated ICT technology and its implementation through system analysis to enable decisions. Particularly, the dynamic linkage between real on-time data and computer analysis is very useful when using GPS, real time sensors, and actuators, such as precision agriculture. The system analysis methods must be changed from static applications to dynamic applications to solve problems over a short time. Further discussions will occur regarding the roles of ICT in sustainability. In addition, indicators of ICT development will be introduced that explain the further development of decision support in agriculture and the quality of life in urban and rural areas.

Chapter 8

SUSTAINABILITY AND AGRICULTURAL INFORMATICS: A GATEWAY FOR ACCESSIBILITY TO INFORMATION SHARING AND THE ICT DEVELOPMENT INDEX

Almost all environmental parameters are connected. For example, food, water, population, urbanization, energy, poverty, health and climate change can be linked together and correlated. Sustainability is the ideal target for connecting parameters that are expected to result in sustainable development. Informatics considers the entire multidisciplinary field of technology transformation along with lifestyle influences. Global demands for food are expected to increase by 60% over the next 37 years if the world's population reaches 9.2 billion by 2050 (FAO, 2013d). Our focus in this chapter is on visualizing the roles of informatics in agricultural sustainability and food security. Sustainable agriculture, food security, and information technology for agricultural purposes are important issues that facilitate modern day reflections regarding the progress of humanity and global stability.

8.1. AGRICULTURE AND INFORMATION TECHNOLOGY

Agriculture is one of the most important sectors in the world, and is not only a food provider but also a source of employment. However, with the development of technology and rapid population growth drived increased food requirements. To meet the demand of food requirements, the agricultural sector must coordinate with other non-agricultural sectors to achieve sustainability. Then, greater agricultural productivity in developed countries would hypothetically create an industrial sector capable of further increases in agricultural productivity in the form of cheaper services. Furthermore, increasing productivity would build the capacity of countries to generate continuous agricultural innovations that would increase the demands for industrial inputs (Hayami and Ruttan, 1971). The greatest concern regarding this hypothesis is the effectiveness of the market information, sufficient research, a proper supply chain, and the availability of markets for products. Therefore, appropriate interactions among these elements and efficient management will lead to success in the technological sector.

In recent years, Asia has faced many challenges in the agricultural sector. The current status of Asia's agricultural sector is being challenged by the demands of the growing population and simultaneous attempts to sustain limited natural resources, respond to climate change and face unstable food prices. Innovations in this sector must deal with all of these challenges and be practical for farmers to implement. The region's population growth is high and it is expected to grow by 142% (South Asia) and 120% (East Asia and Pacific) by 2025 (Population Division, United Nations, 2001). This increasing population must be provided with food, which may only be possible by increasing food production. Simultaneously, agricultural lands are being used by other sectors, mostly for industry and for expanding urban areas. Therefore, large amounts of productive agricultural land have been converted. Furthermore, industrialization and urbanization have resulted in erosion, flooding, drought, salinization, and global warming. All of these factors are causing serious environmental degradation. Concurrently with urbanization, biofuel demand has rapidly increased and large areas of land are being used for palm oil, sunflower, and feed stocks for biofuel production. Consequently, not much arable land is available for agriculture. Numerous governments have taken important steps to protect their agricultural lands and minimize environmental degradation. For instance, laws have been established to distinguish agricultural land from land that is available to other sectors and to protect environmental resources. These laws are strictly imposed. Simultaneously, innovations in the agricultural sector must reach the end user.

Therefore, significant importance should be placed on information technology. In addition to traditional technology, local people should consider improvements to existing technologies. Farmers may experience success with seeds, fertilizers, and pesticides or by using new HYV crops. These experiences are fundamental for incorporating new technology and should be shared globally. Furthermore, other regions of the world can benefit from new technologies. An open source is needed, and farmers must learn how to obtain required information and apply it in their own context. For example, farmers must learn about marketing. In most cases, especially in developing countries, farmers have reduced production or received lower prices for their crops due to lacking market information. Furthermore, farmers must learn about food processing to reduce production waste. Overall, farmers need information to achieve optimum results for each step of plant cultivation, from seed purchasing to product harvest and sales. This information should be easy to obtain. The Internet is the centerpiece of modern information. In addition, the Internet is cheap and accessible throughout the world and could be made available to rural farmers. Currently, the concept of smart farming, which involves sustainable agricultural practices along with a business approach to agriculture, has become the essence of modern agriculture.

However, this concept has been around from the beginning of agricultural production, specifically, since the first crop was sown and farmers wished to sustain or increase their production in next year. The concept of sustainable agriculture integrates this long-term maintenance of natural systems and optimal production with the principals of minimum inputs, adequate income per farming unit, the fulfillment of basic food needs, and the provision of necessities to rural families and communities (Brown et al., 1987). In addition, the government and other stakeholders associated with agriculture require information for planning, improving and delivering their services to farmers using extension services. Digital opportunities are emerging as a positive force for fostering agricultural growth, poverty reduction, and sustainable resource use.

8.2. EMERGING TECHNOLOGIES IN ICT

Technology has played an important role in developing the agricultural industry. Currently, it is possible to grow crops in a desert using agricultural biotechnology (Behera et al., 2014). Using biotechnology, plants have been designed that can survive in drought conditions. Through genetic engineering, scientists have managed to introduce traits into existing genes to make crops resistant to drought and pests (Maheshwari and Kovalchuk, 2014). This technology is being implemented in developing countries to grow cash crops, such as cotton, because genetically engineered cotton plants are pest-resistant, and therefore grow better than traditional cotton plants and have higher yields. Technology has converted farming into an efficient business. Currently, farmers have digitized several processes. For example, a consumer can place an order directly online and the product is transported from the farm to the consumer while it is still fresh. In addition, farmers can now cultivate more than 2 acres because less labor is needed due to the use of planters and harvesters. In agriculture, time and production have the same importance (plant on time, harvest on time and deliver to market on time). In arid places such as deserts, farmers have embraced technologies for crop irrigation (Carlsson et al., 2014). Modern agricultural technology provides an opportunity for a small number of people to grow vast quantities of food and fiber over a short period of time, which is called timeliness. This process helps make products available to markets on time. With modern transportation, consumers in one region can consume fresh carrots from another part of the world on the same day (for example, a carrot that just left a garden in Africa).

Modern transportation technologies allow farmers to easily transport fertilizers and other farm products to their farms, and speed up the daily supply of agricultural products from farm to market. Post-harvest preservation technologies help to keep tomatoes and other perishable crops fresh during transport to the market. These cooling facilities are fitted on food transportation trucks so that crops, such as tomatoes, are fresh upon delivery. Genetically produced plants, such as potatoes, can resist diseases and pests, resulting in high yields and saving the farmer time (Maheshwari and Kovalchuk, 2014). However, debates regarding GM crops are ongoing, and long-term research facilities are needed before GM crops can reach the market. In livestock farming, the price of feed is moderate and affordable for low-income farmers. Most manufactured animal feeds provide extra nutrition to improve the animal's health, which increases the outputs of these animals. In agriculture, the health of an animal determines its output. Most genetically produced animals can produce more milk than traditional animals, which is beneficial for the farmer because it results in high production. Cross-breeding is very effective in animal grazing because cross-bread animals are stronger and more productive. All of these technologies are associated with agricultural information, which will play a large role in agricultural sustainability in the coming decades.

Agricultural information is one of the most important production factors and could lead to improve development. The application of improved knowledge will drive progress and create the best opportunities for sustainable development and poverty reduction. Furthermore, ICTs can greatly improve the accessibility of market information, commodity inputs and consumer trends for famers, which positively affect the quality and quantity of production. Information on marketing, new plant and animal management practices, pests and diseases, transport availability, additional marketing opportunities and the market prices of farm inputs and outputs are very important elements of an efficient and productive economy. Currently, ICTs are driving agricultural progress by enabling an information and knowledge-based approach rather than input-intensive agriculture. Technology used for ICTs can be further classified as follows: modeling, expert systems and GIS, remote sensing, cloud computing, and data analytics (Figure 8.1).

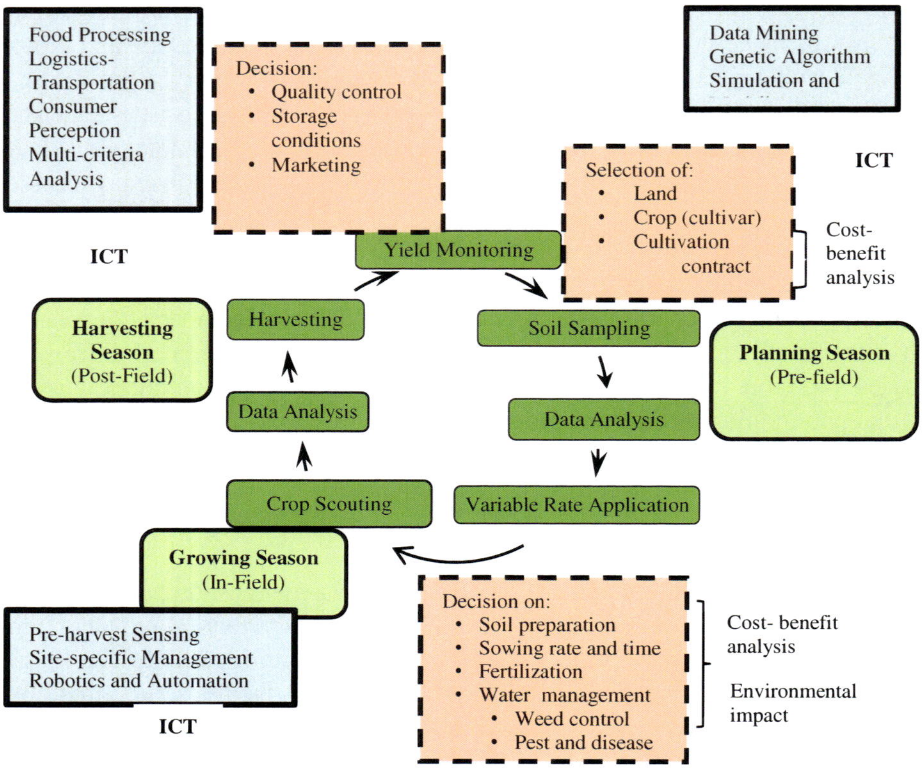

Figure 8.1. Agricultural decision cycles at different stages of production.

Information Technology has become a method for preparing the agricultural sector at local and international levels. Every part of the world is innovating and deploying new technologies to obtain high crop yields to feed the growing population. Some regions are using local methods, while other areas are using new technologies. By contrast, global warming has resulted in profound effects on agriculture. Crop growth patterns and productivity are changing, and excessive use of fertilizer is changing soil properties. In addition, land transformations for urbanization are affecting agriculture. All of these chains of information are very important for the innovations of new technology for agriculture and for updating local farmers. A total mind map of ICT tools has been proposed to obtain the decision of crops during pre-harvest and post-harvest stages (Figure 8.2). Therefore, information technology plays an important role in sustainable agriculture. The main applications of ICT in the agriculture sector from desk to field are briefly described in the following sections.

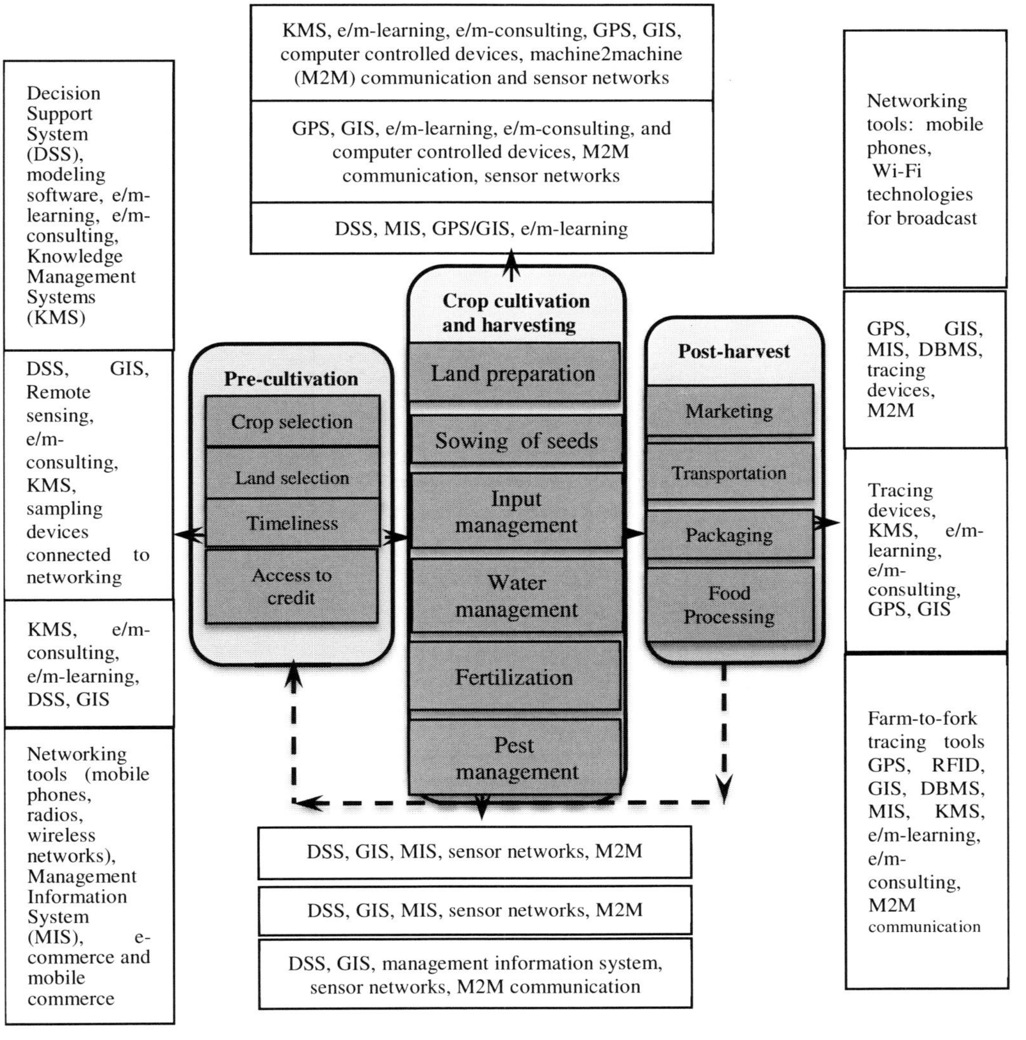

Figure 8.2. ICT tools and technologies for mind-mapping agricultural operations (GFAR, 2009).

8.2.1. Office Automation and Internet

Information technology begins with the use of office automation. Key stakeholders of the agricultural sector, such as government offices, NGOs, research institutes and training centers, should use office automation for data analysis and storage and to improve the efficient development of the agricultural sector. Office automation includes the application of computers, computer networks and telephone networks, which are the core elements of ICT that begin from the desk.

8.2.2. Wireless Technologies

Wireless technologies are used for various purposes in agriculture. The Wireless Sensor Networking (WSN) in agriculture is helpful for collecting distributed data, monitoring in harsh environments, supplying irrigation and fertilizer precisely for crop production while diminishing costs and providing farmers with real time data. One use of wireless technology is its application in simplified plant and yield growth monitoring systems. Currently, smartphones represent the advantages of these technologies. Tablet computers in numerous sizes, especially small ones, are becoming popular, with companies in Asia manufacturing most of them. Given that the ARM processor, which is used in tablet computers, is central to low-cost architecture, a sudden burst of product releases has occurred. The prices of basic tablets have decreased, and processor power and display quality are becoming more advanced. Most tablets can be linked with cell phone networks. Their potential for supporting Agricultural Information Systems (AIS) in the region cannot be underestimated. Mobile communication technologies have become the world's most common method for transmitting voice data and services, and no technology has ever spread faster. At the end of 2010, there were 5.25 billion cellular telephone subscriptions worldwide (Figure 8.3). By 2015, the number of mobile phone connections will likely exceed the global population (Christine et al. 2011). As the use of mobile handsets increases in nearly every country and community, the development of applications will allow us to extend well beyond voice and text communications. Mobile applications for agriculture can provide the most economic, practical, and accessible routes for information, markets, governance, and finance for millions of people who have been unable to use them so far. Significant research has been conducted regarding the development of sensors for monitoring agricultural productivity.

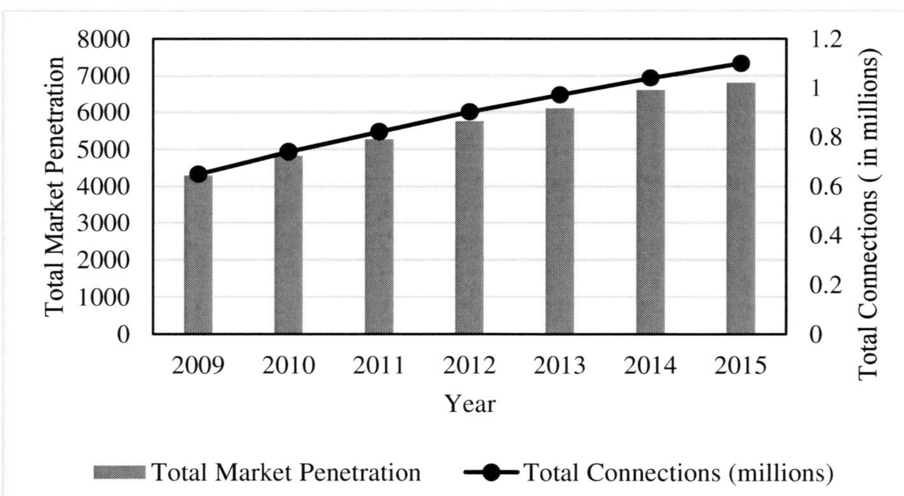

Figure 8.3. Global mobile phone market penetration and connections, 2009-15 (Christine et al., 2011).

8.2.3. Sensors

Sensors that can communicate with each other and with a computer have become more versatile regarding the number of parameters they can cover. Over the last ten years, the price of sensors has considerably declined and their power efficiency factors have drastically increased. Sensors can be deployed extensively for gathering data from remote locations in field-based research or in on-station research. Meteorology, and soil nutrient and moisture management are areas where they can readily be used. In addition, sensors can help detect contaminants and pollutants. Active and passive sensors are readily available on the market for monitoring crop growth and environmental information. When combined with orbital satellite data, sensor-derived information can be used to make accurate forecasts or diagnoses. The use of sensors may be considered essential in agricultural production that is oriented towards global markets.

8.2.4. Global Positioning System (GPS)

In agriculture, GPS is useful for geo-referencing, map-making and surveying. In addition, GPS is used to create various maps related to agriculture and climate change. Using GPS is very simple, and even non-specialists can use it with high accuracy. Today, agricultural equipment uses GPS for sowing, spreading fertilizer, harvesting, and weed collection. Again, observations of earth from space provide a unique opportunity for monitoring the surface mass transfer and deformations related to climate change, particularly when the GPS and Gravity Recovery and Climate Experiment (GRACE) is considered, which can estimate global land and ocean water masses (Jin et al., 2011). Generally, RTK-GPS and GIS applications can be used to develop site-specific management of agricultural inputs to increase farm profitability and reduce environmental risks. Overall, RTK-GPS gathers data with high accuracy (within 1 cm precision), which facilitates precision farming and associated variable-rate technologies. Thus, potential factors that may influence crop productivity can be identified, mapped, and used to promote and implement a solution. This field-level precision allows for better understandings of on-field treatment needs, which were previously considered homogeneous.

8.2.5. Geographic Information System (GIS)

Overall, GIS is a common tool used for monitoring crop yields across seasons to understand the spatial variability of crop yields and identify business opportunities. There are a number of WebGIS applications and tools that are provided on the market. Among them, some of the tools can simulate the growth of specific crops and verify their variations under different climate change scenarios. These tools are usually site-specific but can be applied at national and/or regional levels through a GIS link. The overall output of this type of software is an assessment of crop production under given scenarios, facilitating decision-making from the farm level to the entire cropping system level. Thus, GIS serves as a method for collecting, managing, analyzing, reporting, and sharing vast amounts of agricultural data to help discover and establishing sustainable agriculture practices. A number of tools and software have been developed with the collaboration of research centers and universities throughout the world. For example, The WOFOST, developed by the Centre for World Food Studies and CFWS, in cooperation with the University of Wageningen, can be applied on several different crops, such as barley, field beans, maize, potatoes, rice, soybeans, sunflower, and wheat. The GOSSYM/COMAX model was developed by the Universities of Clemson and Mississippi and the United States Department of Agriculture, and is a combination of the GOSSYM model that is used to simulate cotton growth and COMAX (CrOp MAnagement eXpert), GCMs and weather generators to study the effects of climate change on cotton production. Another good example is the APSIM (Agricultural Production Systems SIMulator), which was developed by a consortium of universities and departments in the Australian state of Queensland (the Agricultural Production Systems Research Unit (APSRU)) and can be applied to more than twenty crops and plants, such as alfalfa, barley, chickpea, cotton, eucalyptus, maize, peanuts, sugarcane, sunflower, tomato, and wheat. The systems coordinate planning and adaptation responses into a set of actions at the farm and regional level, starting with a comprehensive assessment of the impacts of climate change and different farming techniques on crop productivity and agro-ecological system sustainability, and ending with support for the adoption of appropriate agronomic techniques or the establishment of an agro-technology transfer system. Some commonly used decision-support tools include DSSAT (Decision Support System for Agro Technology Transfer), which was developed by the International Consortium for Agricultural Systems Applications (ICASA); CENTURY (from the Natural Resource Ecology Laboratory of Colorado University (NREL)); and MAACV (Model of Agricultural Adaptation to Climatic Variation), which was developed by the Canadian Universities of Guelph and Carleton. Providing the current population and future generations with an indefinite food supply is an economic, environmental, and social priority. Farmers can currently access online agricultural data from government services, including the USDA's soil assessment data or NOAA's weather and climate data, and may integrate these data into their mapping projects. This information allows farmers to make well-informed decisions that increase production and reduce costs with responsible and sustainable practices. The server-based aspect of GIS allows for sharing critical data across the globe and saves valuable time and resources. Mobile ArcGIS technology provides a means of accessing and collecting relevant data in the field for pest management, soil treatment, and weed management (Braun et al., 2008).

8.2.6. Radio-Frequency Identification (RFID)

Previously, RFID has mainly been used in farming. The current application of RFID is for agro-food logistics and supply chain management, animal identification, irrigation, soil temperature measurements, fruit harvesting, and data collection in various environments, such as greenhouses or croplands. In addition, other technologies, such as Knowledge Management Systems (KMS), e-learning, e-commerce, Computer-Aided Manufacturing (CAM), and Computer-Aided Design (CAD) (Ruiz-Garcia and Lunadei, 2011) are used. Three main components constitute RFID systems, a tag or transponder, a reader or transceiver that reads and writes data to a transponder, and a computer containing a database and information management software (Suhong et al., 2006). The RFID tags can be actively, passively, or semi-passively involved. Passive and semi-passive RFID send data through reflection or modulation of the electromagnetic field that is emitted by the reader. The typical reading range is between 10 cm and 3 m. The battery of a semi-passive RFID is only used to power the sensor and for recording data. The RFID has drawn the attention to food logistics and supply chain management processes because of its ability to identify, categorize, and manage the flow of goods (Angeles, 2005). In agriculture, dynamic tags are very interesting, especially for animal behavior studies. These tags automatically send impulses so that even distant readers can identify the animals. This ability is guaranteed by using a power battery, and these devices can be utilized to monitor animals in mid-size outdoor pens to provide digital data that can be easily computerized (Barbari et al., 2008).

8.2.7. Cloud Computing

Cloud computing is a tool for making IT related services available in a simplified manner by hiding their complexities. The users do not need to know who is providing services, and one attraction of cloud computing is that the services are available whenever needed. In addition, cloud computing reduces the costs of obtaining services and offers maintenance. Furthermore, cloud computing creates the network application that is suggested by Google. A narrow goal of cloud computing is to obtain necessary resources through the network. In a broad sense, cloud computing refers to the delivery and use model of services, and allows services to be purchased through the network on-demand. Through cloud computing, which has unique functions, such as ultra-large-scale virtualization and reliability, the providers can handle large amounts of information within a few seconds and achieve the same powerful performance as a super computer (Miao et al., 2013). Cloud computing has universal value for reducing initial cost, allocating resources on demand without limitation, back-end maintenance and upgrades, easy rapid development (including collaboration with other systems in the cloud), and global service development. The core of cloud computing is to manage and schedule information resources by unifying and forming a pool of resources to provide services to users (Figure 8.4). The "cloud" is the network that provides the resource to the users, the resources in the cloud could be unlimited extension or require on-demand payment by volume at any time or anywhere. Therefore, the cloud has some of the following characteristics (Chen et al., 2011).

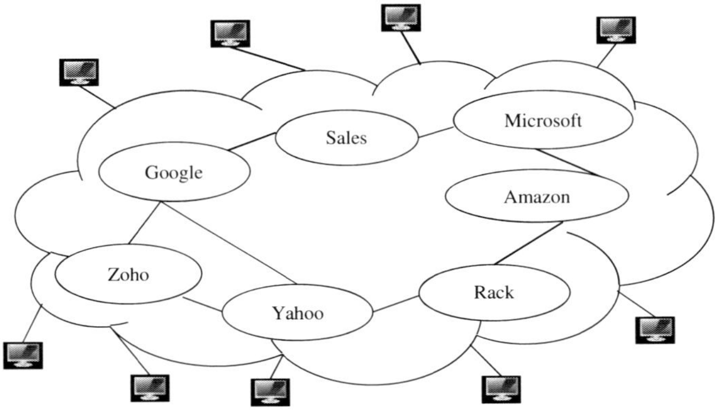

Figure 8.4. Cloud computing (Miao et al., 2013).

8.2.7.1. High Elasticity

Cloud computing includes a massive resource pool that could consist of dynamic stretching according to the applied resource, dynamic expansion of resources when the load is high, and the release of excessive resources when the load is low. Therefore, cloud computing can improve the efficiency of resource use.

8.2.7.2. High Reliability and Flexibility

In the cloud computing process, all data and programs are stored and run in the cloud. In addition, computing is processed by the cloud, cloud services are distributed on the respective servers, and the cloud can automatically handle the failed node and ensure that computing and related applications can run smoothly. For users, this system can be used to quickly take advantage of technology infrastructure resources, the implementation of cloud service mechanisms are transparent, and the users can obtain services that meet their own needs without mastering the cloud computing mechanism.

8.2.7.3. Low Cost, High Sharing and Independence

Because some users may infrequently use the cloud or only use it once, it could be costly. Again, the infrastructure of cloud computing is often provided by a third party and depends on the amount of use, which will greatly reduce the user's cost and the user knowledge requirements. In addition, cloud computing can be used by several users, which prevents individual users from bearing costs that are too high. The users can access the system using a PC or other device and are not constrained by location and time. Instead, users can obtain the information and service over the Internet.

8.2.8. Data Mining

Data mining is the process of extracting important and useful information from large datasets. This information could be converted into useful knowledge that could be used to understand the studied problem and predict future developments. Data mining techniques are largely used in different sectors of the economy and increasingly play a major role in agriculture and environmental-related areas. The idea of using information hidden in relationships among data has inspired researchers in agricultural fields to apply these techniques for predicting future agricultural trends. For example, data collected during wine fermentation can be used to forecast the outcome of fermentation during the early stages of the process. Similarly, soil water parameters for a certain soil type can be estimated using the behaviors of similar soil types (Petraq et al., 2009). Efforts for identifying hidden patterns in data are not new. History shows that extensive data observations have helped identify empirical laws in diverse fields of research. Thus, it is important to provide researchers in agricultural and environmentally related areas with the most advanced knowledge discovery techniques. Recent technologies are currently capable of providing extensive information about agriculturally related activities, which can be analyzed to identify important information. Various data mining techniques have been developed over the years. Some of these techniques are conceptually simple, while others are more complex and may result in the formulation of a global optimization problem. In data mining, data is split into different categories, each of which represents some feature of the data. Several literature reviews have been published regarding data mining techniques. A graphic representation of the classification of data mining techniques is given in Figure 8.5 (Petraq et al., 2009). Data mining methods can be grouped into three categories, statistical, classification, and clustering.

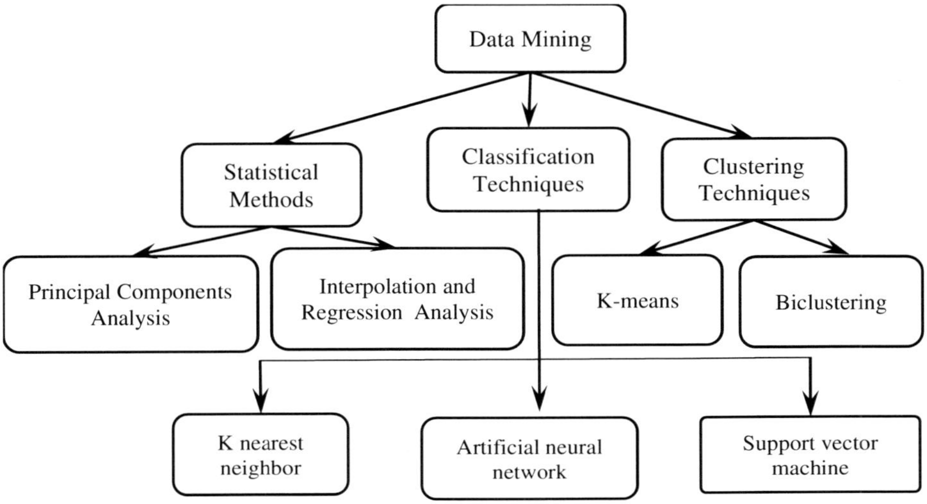

Figure 8.5. Schematic representation of the classification of the data mining techniques (Petraq et al., 2009).

8.2.8.1. Statistical

Statistical methods, such as Principal Component Analysis (PCA), and Interpolation and Regression Techniques (IRT) are commonly used as simple methods for finding patterns in datasets. The PCA method is a useful statistical technique that has been applied in multiple fields, such as image compression, and is commonly used to find patterns in large datasets. The PCA method can help identify patterns in data and express the data in a way that highlights their similarities and differences. Patterns in large data sets can be hard to find because graphical representations of the data are not accessible. Thus, PCA is a powerful tool for analyzing data (Petraq et al., 2009).

8.2.8.2. Classification

Fundamental to the success of any data mining technique is the ability to group available data into disjointed categories, where each category contains data with similar properties. The similarities between different samples are usually measured using a distance function, and similar samples should belong to the same class or cluster of classes. Therefore, the success of a data mining technique depends on the adequate definition of the suitable distance between data samples.

8.2.8.3. Clustering

Clustering techniques are organized into hierarchies and partitioning. The hierarchical clustering approach builds a tree of clusters. The root of this tree can be a cluster that contains all of the data. Then, branch by branch, the initial large cluster is split into sub-clusters until a partition with the desired number of clusters is reached. In this case, the hierarchical clustering is designated as divisive.

8.2.9. Data Mining Applications in Agriculture

Data mining techniques have been extensively used in various fields of study to analyze data by discovering hidden relationships, associations and patterns of data. Information gathered through these techniques can be useful for solving a particular problem. Likewise, in the field of agriculture, data mining techniques are becoming popular because researchers are only interested in extracting useful information hidden in the data to predict future agricultural trends. For example, sets of sounds recorded from animals can reveal the presence of disease or bad environmental conditions. Early detection of these diseases can help farmers take safety measures because sick animals can cause widespread infection. Therefore, a computational system is being development for monitoring pig sounds with microphones installed on the farm and should be able to discriminate among different detected sounds (Chedad et al., 2001). Data collected in the form of images of apples can be used to determine defects in apples, such as water cores, that can change taste and texture. This technique can be used to select good quality apples for marketing purposes. The quality of a fine wine depends on its fermentation process. Therefore, it is very important to predict how fermentation is going during the early stages to ensure good production. Data analyzed though data mining can be used to detect the nature of fermentation to regulate the process. Agricultural pesticides include either chemical or biological agents (such as a virus, bacterium, antimicrobial, or disinfectant) that can deter, incapacitate, kill, or otherwise discourage pests. Although pesticides have benefits, some also have drawbacks, including potential toxicity to humans and other desirable species. Abundant use of pesticides can hamper crop production. Data mining techniques can be used to identify a farmer's use pattern so that steps can be taken to reduce pesticide use. Coarse estimates of recorded cotton pest scouting were made from approximately 1.5 million records and growing (Abdullah et al., 2004). The data indicated farmers' usual patterns of pesticide use on their crops. This research has helped farmers understand suitable timing and accurately determine pesticide application rates for their fields.

8.2.10. Big Data

Big data usually includes datasets that exceed the size-capacity of commonly used software tools to capture, curate, manage, and process the data within a tolerable elapsed time (Economist, 2012). Big data uses inductive statistics and concepts from nonlinear system identification to infer laws (regressions, nonlinear relationships, and causal effects) from large data sets (Delort, 2013), to reveal relationships and dependencies, and to predict outcomes and behaviors (Bittinger, et al., 2005).

8.2.10.1. Big Data to Revolutionize Agriculture

Recent technologies, such as mobile sensing devices, aerial sensory technologies (remote sensing), software logs, cameras, microphones, and RFID readers, provide abundant amounts of data regarding agriculturally related activities wirelessly. Today's farmers have extensive field data. However, it is difficult for farmers to know how to use the data for decision-making. This scenario has created an excessive need for collecting, storing and analyze data to obtain sufficient information. Therefore, researchers and scientist are focusing more on big data. Big data can be helpful in agricultural research because more data can lead to analysis that is more accurate. For example, the development and production of feed plants and animals are important components of food security. The few notable research activities that have been conducted regarding the computational biology of plants and animals and their pathogens have resulted in a transformation that has enabled scientists and organizations to better feed the world and improve the qualities of food crops and livestock (Ryan, 2014). In addition, these disciplines involve very large datasets and require computationally expensive analyses. Big data that uses computational analysis can help agriculturalists design, track and analyze their farming operations through an entire farm cycle as efficiently as possible.

With the help of modern science, many agricultural technology innovations have been developed and many are in the process of development. Before applying these technologies, experiments, discussions and pilot tests must be conducted. In addition, some technology must undergo further development to use in a different region. Technologies that require minimal application costs are expected to be adopted widely. It is important for non-specialists to adopt this technology to obtain the best use of technology. Here, the core element of adaptation is information technology and how technology reaches the user.

However, one concern is how much these technologies are adopted by the user. In most cases, new technology cannot reach its expected outcome because it is not user friendly. In most cases, farmers try to use conventional techniques rather than adopt new techniques. Researchers always try to convince farmers to adopt new technology by assuring them that they will benefit from it. However, the famers do not take the technology further in many cases because the results do not meet their ultimate goals. Therefore, researchers must focus on understanding why farmers do not want to use new technology, and how to adopt it and the required timeframe for adapting to modern technology. Several theoretical models explain the speed and rate of the adaptation process, which are known as diffusion of innovations models.

8.3. Diffusion of Innovation Theory (DOI)

Diffusion of innovation (DOI) is a theory that aims to explain how, why, and at what rate new ideas and technologies spread through cultures. Innovation adoption is an important element for sustaining modern technology over the long term. In addition, DOI plays an important role in agricultural sustainability. Diffusion is a process by which an innovation is shared in different ways over a period among individuals in a social system. An innovation is an idea, practice, or object that is considered new to an individual or other adopting entity. Communication is a process by which participants create and exchange information with one another to reach a mutual understanding (Rogers, 1995). Studies have shown significant differences in the characteristics of early innovation adopters and late innovation adopters. However, one must know the basic characteristics of the people who would be the ultimate innovation user. The adoption process must be applied to a category of users. An overview of the five basic adopter categories according to Rogers's research (1995) is given below.

Innovators – There is always an enthusiastic group of people who want to be the first to use a new innovation. These people wish to take a risk, try something new and are willing to contribute their ideas and experiences throughout the process to develop the technology. The introducers of the technology would not have to persuade them much to use the technology.

Early Adopters – This group holds knowledge regarding the need for the innovation and tries to take leadership roles in introducing the innovation to their own community. These individuals are comfortable considering new technology and only require that the technology is user friendly with manuals and implementation information.

Early Majority- These people always want to be the first to know about practical success stories featuring an innovation. Unless they have knowledge of positive outputs from adopting a new technology, it will be very hard to convince them to use it.

Late Majority- This group of people only adopts innovations when they are widely adopted by other people. These people are interested in knowing how many people have tried the innovation and want feedback regarding its success.

Laggards- This group is the most challenging group to convince to use new technology. This group is fully dependent on conventional techniques and hardly ever tries new techniques. Apart from a successful trial, this group may only be convinced when they are forced to do it by other adopter groups.

There are five basic characteristics (relative advantage, compatibility, complexity, trialability, and observability) that inspire adaptation of innovations. These characteristics work in different ways in the five adopter categories. However, one relative advantage has been defined as the degree to which an innovation is evaluated as better than an existing idea, process or product. Compatibility is measured based on how consistent the innovation is based on experiences, values and the requirements of the potential adopters. Complexity is the most important characteristic because it measures how difficult the innovation is to understand or use most people. Reliability indicates the extent to which the innovation can be experimented with before the user must commit to adopting it. Observability is the extent to which the adoption is observable to people. The more visible the adoption, the more it will spread to the majority of people and obtain positive or negative feedback from users. The above-mentioned five stages of diffusion theory are widely applied in information technology to understand how a technology is expected to spread and the key factors that will inhibit the adoption of the new technology by users. Therefore, innovation diffusion plays an important role in the agricultural sector. Many innovations are coming, but few are equally available. Cost or farmer efficiency may be not be the only reasons for adopting new technology.

8.4. Technology Acceptance Model (TAM)

Technology acceptance is an information systems theory model that shows how users accept and use a technology. This model suggests that a number of factors influence user decisions regarding how and when to use a new technology that they are presented with. Thus, the perceived usefulness (PU), which is defined by Fred Davis as the degree by which a person believes that using a particular system would enhance his or her job performance, and the Perceived Ease-of-Use (PEOU), which is the degree to which a person believes that using a particular system would not require effort, are considered (Davis, 1989).

8.5. Business Model of Food Security

Information and knowledge play important roles in ensuring food security and sustainable development in agriculture. In the Chapter 4, food security concepts and dimensions are discussed throughly. In summary, it is noted that food security is achieved when all people have physical and economic access to sufficient, safe, and nutritious food at all times to meet their dietary needs and food preferences to lead an active and healthy life. Food security depends on food availability, access, utilization and the stability of the supply and is linked to a secure livelihood. Food security assessments include indicators of food availability, access and nutritional status. When a population with adequate access to food is affected by malnutrition, the dietary diversity, nutrient intake and diet affordability should be assessed because people require "Food and Nutrient Security". Food security and safety are tightly linked and different. Food safety is a scientific discipline that describes the handling, preparation, and storage of food in ways that prevent food-borne illness. If food is contaminated and makes people sick, it detracts from food security. As ICTs are developed and spread throughout the world, opportunities for broadening rural development will arise. However, technology may not be accessible for an entire region, particularly in developing countries due to financial and social barriers. This limitation must be minimized for an individual rural area by adopting and practicing new technologies that could be adopted with minimal resources and knowledge. The first important element is 'access'. Access could be explained as a method for making data available to all users, which would allow all users the opportunity to enter into a network of information in a communication system through the Internet. This is an important element for policy discussions and a requirement in all development activities. The main purpose of access is to share information with all consumers about their daily affairs related to economic, social, cultural, and political life. As part of the other communication media, this data helps them visualize facts and increases their creativity. This network service is for all people throughout the world. However, one element of networking is to recognize the obstacles for accessing a network of individuals of various groups distinguished by age, income, education, language, and location. Recent activity, inventions, technology, research, publication, interviews, and problem analyses should be included. The same results may be found in other conventional media, but the difference is the simplicity and ease of understanding for users. In this regard, Clement and Shade (2000) developed a model that is known as 'Access Rainbow'. This model illustrates the multifaceted nature of access. In this model, the lower layers emphasize the conventional technical aspects and are complemented by the upper layers. However, all of the layers are linked to each other and together give proper access.

8.5.1 Accessing Information to Develop a Business Model

8.5.1.1. Carriage Facilities

Carriage facilities result in the highest possible infrastructure for transporting data with minimum cost to the end users. However, these facilities would not be located in rural areas. This becomes an important concern where everyone is trying to arrange a facility within limited resources. Computers and other technical devices are too expensive for lower income people. Furthermore, many devices are not user friendly for everyone. A browser, email and other software are needed to use the Internet. Although software significantly expands the facility of ICT, it is expensive to buy. This software provides wide access to online databases and website information sources, such as email and e-commerce services, without charge to the user. Local ISPs (Internet service providers), community access points, text and computer literacy, training and support services, and other social facilitation are required. Furthermore, good governance and public consultation regarding policy issues and social impact assessments are needed.

8.5.1.2. Telecentres

Telecentres are one type of public access service that is a centerpiece for communication and supports small businesses in rural areas. Their services include the following: (1) providing information about crops, weather, market information, local production support services, fertilizers, and online production counseling; (2) educated employees can be asked to translate necessary information for local people; (3) the telecentres may extend their services by opening an index website and a search engine to meet the specific requirements of local people.

8.5.2. Community Informatics

Community Informatics (CI) is a concept that connects the people in a community to the ICT. The ultimate objective of CI is to achieve the community's goal of economic development and to support cultural affairs and environmental protection. Special concern is given to developing countries so that every individual has the opportunity to access the ICT if it is available. Therefore, the organization through which community access is provided plays an essential role in optimizing its function. In addition, CI is an approach to ICT that includes concern regarding the accessibility of hardware and software and the connectivity of information. Furthermore, CI evaluates the user by which the technology is being applied and the technology use in the context of the user's physical community. Incorporating the user and his community into the system design process introduces new elements and different "stakeholders" into an extended approach to ICT design, development, and implementation (Michael Gurstein, 2000). Access facilities are an important concern for the community informatics approach, which provides users access to modern technologies, especially in less connected communities, through telecentres, CAP sites, and cyber cafes. A design service should be provided to segregate and accumulate the information needed by an individual community. Telecentres are central to rural development in the ICT context and should be designed to work effectively as information providers and users. Furthermore, telecentres should work with intentional linkages and bases, provide connectivity, include multiple uses for education and small business support, and provide communication and language translation to the community. Furthermore, many services, such as information, education and training, mentoring and consultation, diagnosis and monitoring, and transaction processing are providing through the Internet to the local people and should be supported. In this regard, a'Community Toolbox' serves as an excellent example. The Community Tool Box is part of the University of Kansas, KU Work Group's role as a designated World Health Organization Collaborating Centre for Community Health and Development (Holt et al., 2013). The

Community Toolbox offers free online services to help people who need information regarding community assessment, planning, intervention, evaluation, advocacy, and other community services. Furthermore, the Community Toolbox is involved in teaching, training and technical support. This service has spread to over 230 countries around the world.

8.5.3. e-Commerce

Electronic commerce, commonly known as e-commerce, is used for buying and selling products or services using electronic systems, such as the Internet and networks. E-Commerce has assumed an important role in agri-business and is contributing to meeting individual requirements and needs. Using e-commerce, a seller can find a buyer before harvesting. Transactions have become easy and cost effective through this system. Agricultural marketing and the distribution of information to farmers not only helps farmers sell at good prices, but also helps communicate reliable food price information to policy makers. To widen these affects, many commercial and non-commercial agencies are now working to spread e-commerce to rural communities in developing countries to link local people directly to retail, trade purchasers and suppliers. In this way, the need for mediators in the supply chain from farmer to consumer is reduced. In addition, this method will allow farmers to obtain their expected price for their hard work.

8.5.4. e-Learning

Currently, the Internet and other applications offer numerous advantages. We can obtain required information quickly from any corner of the world. Each age group, whether a child or homemaker, young person or senior citizen, is now using the Internet for his or her own purposes. Students, researchers, scientists and the public can easily access journals, papers, and books. Businessmen can advertise their products using e-commerce. A consumer can buy any product through the Internet. One example of Internet technology is e-learning. E-learning has become a general term that encompasses the application of computer technologies to education, whether face to face or in classrooms, in blended and hybrid courses, in a mediated distance education context or in online learning environments (Abrami et al., 2006). Meanwhile, Stockley (2003) noted that e-learning is the digital delivery of a knowledge, training or education program. E-learning involves using a computer or electronic device for some method of training, education or learning of material. Similarly, e-learning could be very useful for corporate activities. For example, in organizations, training is an integral program for all departments, from new employees to senior employees, Everyone must go through training programs regarding various subjects. In such cases, proper time management would not allow for classroom training. In this case, e-learning could be more efficient because employees could attain training at their convenience. Currently, students, teachers, scholars and/or specialists of various disciplines have given lectures through the Internet. In addition, the public can learn from these sources without needing a classroom, and the sources are free. Therefore, e-learning could be very helpful to the agricultural sector. In fact, some agricultural universities in Canada, UK, and the United States provide online degrees in agricultural related subjects so that anyone, anywhere and anytime, can attend online classes and take exams. However, in other parts of world, especially in developing countries, these facilities are limited or have not been introduced. Therefore, the FAO, along with collaborating partners, have organized e-learning for key people in the agricultural sector, including researchers, policy makers, and scientists. Several important e-learning programs are discussed briefly.

The "Information Management Resource Kit (IMARK)" attempted to design a curriculum to provide necessary information to professionals in various important areas of agricultural information management (Nadeau and Rudgard, 2007). The most important feature of IMARK is that it offers modules in several languages. In addition, "Food Security Information for Action" is a curriculum designed by the FSNAU (Food Security and Nutrition Analysis Unit) in collaboration with the EC/FAO. This program includes a series of eLearning courses for learners who are associated with food security related issues, such as information collection, management, data analysis and reporting. The third example is "A Primer to the Right to Adequate Food" which is part of the e-learning curriculum "Right to Food in Practice". This program focuses on the core issues behind the human right to access adequate food. In addition, this program defines the rights, commitments and responsibilities of rights-holders and provides guidelines. This course is primarily for UN country teams and FAO staff, specialists, NGOs/CSOs, the public, and academic institutions. Another program, "Enhancing Participation in Codex Activities", is designed for professionals who are involved in setting food standards and food control mechanisms. This course provides a comprehensive description of the organization of the Codex Alimentarius Commission (CAC), including its operation and management (FAO and WHO, 2005). These courses are especially designed to fulfill National Codex Contact Points (NCCP) and be used by government officials, representatives of the food industry, consumer groups and Codex observer organizations.

Undoubtedly, information technology can play a major role in agricultural sustainability. Therefore, information technology is needed to study global and local contexts. This broad sector should be disseminated throughout the world. Technology may be expensive for an end user, but information technology can be adopted and practiced with minimum cost. The above e-learning examples and the development of Asia's e-learning sector indicate that sustainable development could be achieved if information is easily shared, gathered, and implemented.

8.5.5. e-Sourcebook

The ICT in Agriculture e-Sourcebook, a codex of e-learning, was designed by the United Nations Food and Agriculture Organization. This e-book was developed jointly by the World Bank's Agricultural and Rural Development Sector and infoDev, and it has benefited from generous funding from the Government of Finland under the Finland/infoDev/Nokia program "Creating Sustainable Businesses in the Knowledge Economy." This e-book is designed to support practitioners and policy makers in taking advantage of the potential of ICTs as tools for improving agricultural productivity and smallholder incomes, strengthening agricultural markets and institutions, improving agricultural services, and building developing-country linkages to regional and global agricultural value chains (McNamara et al., 2011). The e-sourcebook focuses primarily on how ICT can assist small-scale producers and the intermediate institutions that serve them. In addition, the e-sourcebook considers how to link smallholders to ICT-enabled improvements in larger-scale farming, markets, and agribusiness to stimulate the broader rural economy. A comprehensive overview of current and upcoming ICT-in-agriculture applications and how they might improve agricultural interventions or strategies can be easily learned from the e-sourcebook.

8.5.6. Mobile Technology

Mobile communications technology has become the world's most common method for transmitting voice, data, and services, and no technology has ever spread faster. With increased use of mobile handsets in nearly every country and community, applications have been developed for uses that extend well beyond voice and text communications. Mobile applications for agriculture can provide economic, practical, and accessible routes to information, markets, governance, and finance for millions of people who have been excluded from their use. M-apps software is designed to take advantage of mobile technology and enable the collection and transmission of data for economic and social activities, whether for commercial, administrative, or entertainment purposes (McNamara et al., 2011). Even by 2015, standard mobile phones will outnumber smart phones in developing countries by nearly four to one (Figure 8.6). Moreover, m-apps do not need to be associated with specific access devices, but rather focus on providing information and facilitating activities. Touch devices are used at fixed locations and can provide more information though their monitors than the smaller screens on mobile phones. However, mobile phones have several advantages over less mobile (or fixed location) devices, such as touch screens. Furthermore, most m-apps can be replicated across different mobile interfaces and devices, such as SMS phones, mobile browsers, smart phones, and tablets.

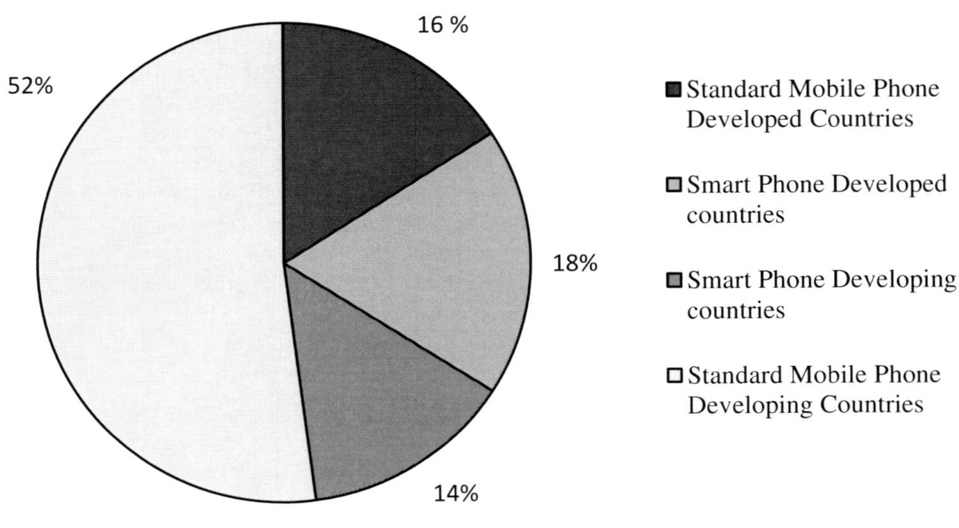

Figure 8.6. Forecast of global mobile phone use by type in 2015 (Christine et al., 2011).

Table 8.1 Common Apps Used for Agriculture and Weather

Google Earth®	Google Earth allows users to search by voice for cities, places and businesses. Users can browse layers, including roads, borders, places, photos and more. Google Earth is a mapping application that operates on Apple and Android devices.
ArcGIS®	ArcGIS helps to query maps, find interesting information, measure distances and areas of interest and share maps with others. It is a mapping application that operates on Windows, Apple and Android devices.
Pocket Weather®	Pocket Weather provides regular updates on weather and daily and weekly views.
Oz weather®	Oz weather provides frequently updated weather forecasts, observations and radar loops from the official Australian Bureau of Meteorology. More than 250 Australian town and city forecast locations are available in this application. It is a weather application and is most commonly used on Apple devices.
F-Track Live®	A farm management app that allows multiple users record and access farm information, such as stock records, crops, and tasks, and reference information in real time. In addition, F-Track Live replaces the need for a paper-based stock book and is used in the form of a Farm Management system that operates on Apple and Android devices.
Mobile Pioneer®	Mobile Pioneer is designed to provide information about local weather and markets, field record keeping, agronomy resources and more. It is in the form of a Farm Management system and operates on Android devices.
CaleWelator®	Lambing or just losing count of all those sheep? Easily record them and see your lambing stats with the handy farm app CaleWelator. CaleWelator is a stock-management app and uses the Android interface.
Agvantage®	This app is designed as an easy, quick and portable method for determining the dose rate for sheep and cattle for a range of animal health treatments. This Apple–based stock management app is used in mobile applications.
Cropmate Variety Chooser®	Crop Mate Variety Chooser is a decision support tool for farmers to help them choose varieties of barley, canola, chickpeas, oats, triticale and wheat. This tool has cropping and horticultural applications for smart phone users.
Hort Service Finder®	The Horticulture Service Finder application is provided to users as a free service to help Australian horticultural growers to find horticultural service providers near them, including government and private agronomists, financial service providers and more. This application is also for iPhone users.

Innovation is needed and can be achieved through more effective products, processes, services, technologies or ideas. The need to feed the growing planet requires us to increase our output with available resources and IT, which offers many opportunities for addressing challenges to increase global agricultural output and must play a central role in meeting the goal of food security. Recently, ICTs have played an important role in promoting innovation in the agricultural sector. The values of these innovative technologies and services should not be underestimated because improving agricultural extension services to farmers using mobile technology would effectively improve the transmission of agricultural research results for application in fields (Table 8.1). Although IT has influenced many industries and has greatly improved the management of supply chains, it has only influenced agriculture in a relatively localized way. In the next section, the sustainability index for ICTs is discussed in the context of ICT development indicators. A few common indicators were reviewed related to the ICT infrastructure and Access Index:

8.6. ICT DEVELOPMENT INDICATORS

To measure ICT development indicators, an ICT development Index IDI was introduced at the beginning with sets of 11 indicators in access, use and skill categories with one benchmark measure for monitoring and comparing the developments in ICT across the countries (ITU, 2012). Later, the ICT opportunity index (ICT-OI) was used to measure the ICT application of Data Envelopment Analysis (Emrouznejad et al., 2010). The ICT-OI has generally been accepted as a statistical tool for tracking digital device by measuring the relative difference in ICT opportunity levels among economies and regions over time. The ICT-OI is based on the dual concepts of a country's productive capacity and consumption. Therefore, the ICT-OI depends on the degree of a country's infodensity (the slice of a country's overall capital and labor stocks, indicative of its productive capacity) and info-use (the consumption flows of ICT). Technically, the ICT-opportunity value is derived and expressed as follows:

$$\text{ICT-OI} = \sqrt[2]{(info - density\, x\, info - use)} \qquad (8\text{-}1)$$

The ICT-OI consists of 10 indicators that are grouped into 4 sub-indices, networks and skills as indicators of infodensity, and uptake and intensity as measures of info-use. Networks are captured as the penetration rates of fixed telephone lines, mobile cellular subscribers, and international Internet bandwidth. Skills are measured by adult literacy rates and gross enrolment rates for primary, secondary, and tertiary education. Uptake is captured through Internet users, computers per 100 inhabitants, and the proportion of households with a TV. Intensity indicators include broadband Internet subscribers per 100 inhabitants and international outgoing telephone traffic (min) per capita. The 2007 ICT-OI splits the world economies into four major groups depending on the degree to which the country has access to and uses ICT (high levels (249 and above), upper levels (between 150 and 248), medium levels (between 68 and 149), and low levels (between 12 and 68)). Data envelopment analysis was conducted as a linear programming technique to assess production management. The Data Envelopment Analysis (DEA) is a linear programming approach and was analyzed as an output-oriented case. The DAE method seeks a maximum proportionate increase in output production based on input levels. For example, an output-oriented model that is considered under the assumption of variable returns to scale (VRS) can be expressed to maximize α as follows:

$$\sum_{j=1}^{n} \lambda_j X_{ij} \leq X_{ij0} \qquad ; i = 1, \dots \dots, m \qquad (8\text{-}2)$$

$$\sum_{j=1}^{n} \lambda_j Y_{rj} \geq \alpha Y_{rj0} \quad ; r = 1, \dots \dots, s \tag{8-3}$$

$$\sum_{j=1}^{n} \lambda_j = 1. \ \lambda_j \geq 0 \quad ; j = 1, \dots \dots, n \tag{8-4}$$

with an output-oriented DEA model with input variables ($X_1, \ldots, X_m$),output variables ($Y_1, \ldots, Y_s$), and with n decision making units (j = 1, ..., n).

8.7. ICT Infrastructure and Access Index

The ICT infrastructure and access growth can be measured to determine sustainability over time by using 12 growth indicators (Hanafizadeh et al., 2009). The major indicators include the following: fixed telephones, mobile cellular inhabitants, computers, Internet subscribers per 100 inhabitants and population density covered by mobile cellular service. The other two indicators were considered Internet access tariffs and mobile cellular tariffs. Indicators that were selected across the country or had regional ICT infrastructure analyses required access aggregation and index construction with different units. Normalization is useful for creating a common unit among indicators. A number of normalization methods are available that exploit standardized methods using the following equation:

$$Iqc = \frac{Xqc - Xqc = \bar{c}}{\delta qc = \bar{c}} \tag{8-5}$$

where Xqc is the numerical value of the qth basic indicator in the cth country, $Xqc = \bar{c}$is the average of the countries' numerical values in the qth indicator, and $\delta qc = \bar{c}$ is the standard deviation of the qth indicator among the countries. The normalized weights of each indicator and in each factor is the squared factor loading of that indicator divided by the sum of squares factor of the indicator loadings on the factor after rotation (Nicoletti et al., 2000). This value, which is the variance portion of the factor, is explained by the following indicator:

$$w_{ij}^{k} = \frac{(l_{ij}^{k})2}{\sum_i (l_{ij}^{k})2} \text{for} \begin{cases} j = 1; \ k = 1, 2, \dots, 7 \\ j = 2; \ i, k = 1, 2, 3 \end{cases} \tag{8-6}$$

where w_{ij}^{k} is the weight of the ith indicator (that lies out in the jth dimension) for the kth factor and l_{ij}^{k} is the factor loading of the ith indicator (that lies out in j^{th} dimension) for the kth factor.

$$F_{jc}^{k} = \sum_i w_{ij}^{k} \ I_{ijc} \ for \begin{cases} j = 1; \quad k = 1, 2, \dots., 7 \\ j = 2; \quad k = 1, 2, 3 \\ c = 1, 2 \dots.., 150 \end{cases} \tag{8-7}$$

Here, I_{ijc} is the value of the ith indicator (that lies out in j^{th} dimension) for the cth country, and the F_{jc}^{k} is the value of the kth factor for the cth country in the jth dimension. Then, a linear aggregation of the factor values in each dimension is calculated as follows:

$$CI_{jc} = \sum_k w_j^k F_{jc}^k \ for \begin{cases} j = 1; & k = 1,2,\ldots.,7 \\ j = 2; & k = 1,2,3 \\ c = 1,2\ldots..,150 \end{cases} \quad (8\text{-}8)$$

where CI_{jc} is the value of the jth dimension index for the cth country, and w^k_j is the weight of the kth factor in the jth dimension. It is noteworthy that the weight of each factor in each dimension is the variance of that dimension as explained by the factor and can be expressed as follows:

$$w_j^k = \frac{\sum_i (l_{ij}^k)2}{\sum_i (l_{ij}^k)2} \text{for} \begin{cases} j = 1; \ k = 1,2,\ldots,7 \\ j = 2; \ i,k = 1,2,3 \end{cases} \quad (8\text{-}9)$$

The selection and normalization of indicators, the aggregation model, and the weights of the indicators, greatly influence the sustainability assessment of the ICT indexes regionally. The price of the ICT services significantly affects the demand and spread of the ICTs for sustainability. To monitor the affordability of ICT services across the region and over time, the International Telecommunication Union (ITU) developed the ICT Price Basket (IPB), which is a comparison tool that provides insightful information for determining the costs and affordability of fixed-telephone, mobile-cellular, and fixed-broadband services (ITU, 2012). Information could result in the power to require ICT infrastructure for sustainability purposes and result in access by households and individuals. The ICT Development Index and IPB are top benchmarks for predicting uses and ICT skills development in society. In addition, ICT knowledge contributes data that are very meaningful when they are implemented through knowledge and business. Our big data discussion highlights the fact that database value creation requires the identification of patterns from which predictions can be inferred and decisions can be made. Businesses must decide which data to use (World Economic Forum, 2014). The data owned by each business could be as different as the businesses themselves and range from log files and GPS data to customer-to-machine or machine-to-machine data. Building a network that maximizes the impacts of Big Data requires powerful and seamless interactions among sensors, devices, computing, storage, analytics, and control systems.

REFERENCES

Abdullah, A., Brobst, S., Pervaiz, I., Umar, M. and Nisar, A. (2004), Learning Dynamics of Pesticide Abuse through Data Mining, Australasian Workshop on Data Mining and Web Intelligence, Dunedin, New Zealand.

Abrami, P. et al., (2006), A Reviews of e-learning in Canada: A Rough Sketch of the Evidence, Gaps and Promising Directions, *Canadian Journal of Learning and Technology,* 32(3): Fall.

Afgan, N. H. and Carvalho, M. G. (2004), Sustainability Assessment of Hydrogen Energy Systems, *Int. J. Hydrogen Energy,* 29:1327–1342.

Ahamed, T., Takigawa, T., Ryozo, N. and Tian, L. (2014), Bioproduction Engineering- A Road Map for Sustainable Agricultural Practices, *Nova Science Publishers Inc, USA.*

Alam, M. S., Bala, B. K. and Huq, M. Z. (1997), Simulation of Integrated Rural Energy System for Farming in Bangladesh, Energy, 22(6): 591-599.

Angeles, R. (2005), RFID Technologies: Supply-chain Applications and Implementation Issues, *Inform. Syst. Manage,* 22:51–65.

Arizpe, L., Constanza, R. and Lutz, W. (1991), Primary Factors Affecting Population and Natural Resource Use, International Conference on an Agenda of Science for Environment and Development into the 21th Century, Vienna.

Barbari, M., Conti, L. and Simonini, S. (2008), Spatial Identification of Animals in Different Breeding Systems to Monitor Behavior. In: ASABE, Editor, Livestock Environment VIII. Iguassu Falls, Brazil.

Bartoszezuk, P., Ma, T., Nakamari, Y., (2001), Environmental Kuznets Curve for Some Countries—regression and agent-based approach. Japan Advance Institute of Science and Technology (mimeo).

Beddington, et al., (2009), Achieving Food Security in Face of Climate Change, Commission on Sustainable Agriculture and Climate Change, 1-18.

Behera, B., Singdevsachan, S., Mishra, R., Dutta, S. and Thatoi, H. (2014), Diversity, Mechanism and Biotechnology of Phosphate Solubilizing Microorganism in Mangrovea Review, Biocatalysis and Agricultural Biotechnology, 3(2): 97 – 110.

Bell, S. (1996a), Approaches to Participatory Monitoring and Evaluation in Dir: District, North West Frontier Province, Pakistan, Systems Practice, 9(2):129–150.

Bell, S. (1996b), *Learning with Information Systems: Learning Cycles in Information Systems Development,* Routledge, London, UK.

Bell, S. and Morse, S. (2008), Sustainability Indicators, Measuring the Immeasurable? *Earthscan,* London, UK.

Benton, T., Juliet, M., Vickery, A. and Jeremy, D. (2003), Wilson Farmland Biodiversity is Habitat Heterogeneity the key? *TRENDS in Ecology and Evolution,* 18(4).

Biomass Nippon Integrated Strategy (2006).

Biomass Research (2013), Analyzing The Effect of Biofuel Expansion on Land Use in Major Producing Countries, Evidence of Increased Multiple Cropping, Biomass Research Report 1301, Wageningen, Netherland.

Bittinger, M. L., Neal, B. and John, Q. (2005), Calculus for the Life Sciences, *Pearson Addison Willy.*

Boulanger, P. M. and Brechat, T. (2005), Models for Policy Making in Sustainable Development: The state of the art and Perspective for Research, Ecological Economics, 55: 337– 350.

Braun, M., Andrew, P. and Nancy, E. (2008), GIS for Sustainable Agriculture.

Brown, et al., (1987), Global sustainability: Toward Definition, Environmental Management, 11 (6): 713-719.

Cao, K., Huaang, B., Wang, S. and Lin, H. (2012), Sustainable Land Use Optimization Using Boundary-Based Genetic Algorithm, Computers, Environment and Urban Systems, 36: 257-269,

Carlsson, A. S., Zhu, L. H., Andersson, M., and Hofvander, P. (2014), Platform Crops Amenable to Genetic Engineering a Requirement for Successful Production of Bioindustrial Oils Through Genetic Engineering, Biocatalysis and Agricultural Biotechnology, 3(1): 58 – 64.

Carpenter, S. R., Mooney, H. A., Agard, J., Capistrano, D., Fries, R. S., Diaz, S., Dietz, T., Duraiappah, A. K. and Oteng, Y. A. (2009), *Environmental Science and Policy,* 12:373 – 377.

Castro, J., and D'Agnes L. (2008), Fishing for families: Reproductive health and integrated coastal management in the Philippines (Focus Issue 15)." *Washington, DC: Environmental Change and Security Program, Woodrow Wilson Center. Focus, Issue 15.*

CES (2000), Center for Environmental Strategy, Index of Sustainable and Economic Welfare.

Chambers, N., Simmons, C. and Wackernagel, M. (2000) Sharing Nature's Interest: Ecological Footprints as an Indicator of Sustainability, Earth scan, London, UK.

Chambers, R. (1997), Whose Reality Counts? Putting the First Last, Intermediate Technology, London, UK.

Checkland, P. and Holwell, S. (1998), Information, Systems and Information Systems: Making Sense of the Field, *Wiley,* Chichester.

Chedad, A., Moshou, D., Aerts, J. M., Van Hirtu, A., Ramon, H. and Berckmans, D. (2001), Recognition System for Pig Cough based on Probabilistic Neural Networks. *Journal of Agricultural Engineering Research,*79(4): 449–457.

Chen, M., Wang, J. and Niu, Y. (2011), Research on Agricultural Information Resources Sharing System Construction Based on Cloud Computing, *Agric. Network Inform.,* 11(4).

Chisti, Y. (2007) Biotechnology Advances, 25 (3): 294-306.

Christine, Z. Q., Siou, C. K., Andrew, D. and Steve, E. (2011), Mobile Applications for Agriculture and Rural Development, ICT Sector Unit World Bank, Washington DC, USA.

CIGR (1999), CIGR Handbook of Agricultural Engineering, 5, Energy & Biomass Engineering, ASABE USA, 13-21.

Clement and L. Shade (2000), The Access Rainbow: Conceptualizing Universal Access to the Information Communications Infrastructure, Idea Group Publishing, 1-20.

Commission of the European Communities (1984), The Regions of Europe: Second periodic report on the social and economic situation of the regions of the Community, together with a statement of the regional policy committee, OPOCE, Luxembourg.

Corral, L., Winters, P. and Gordillo, G. (2000), Food Insecurity and Vulnerability in Latin America and the Caribbean, Working Paper Series in Agricultural and Resource Economics, University of New England, Australia, 1-31.

Craig, D. and Porter, D. (1997) 'Framing Participation: Development Projects, Professionals, and Organizations', *Development in Practice,* 7(3): 229–236

Craig, J. and Charles, P. (1977), Insurgency of the Powerless: Farm Worker Movements (1946-1972), *American Sociological Review,* 42(2): 249-268.

Dalton, H. and Brand, R. (2003), Nitrogen: the essential public enemy. *Journal of Applied Ecology,*40(5): 771–781.

David, S. (1989), Sustainable Development: theoretical construction attainable goal? *Environment Conservation*, 16: 41–48.

Davis, F. D. (1989), Perceived Usefulness, Perceived Case of Use and User Acceptance of Information Technology, MIS Quarterly, 13(3): 319-340.

Delort, P. (2013), Big Data in Biosciences, Big Data Paris, http://www.big dataparis.com/documents/Pierre-Delort-INSERM.pdf#page=5

Dempster, J. P. (1975), Animal Population Ecology, *Academic Press*, London, UK.

Devereux, S. (2000), Famine in the Twentieth Century, Working Papers, Brighton, Institute of Development Studies, University of Sussex, UK.

Devereux, S. and Maxwell, S. (2001), Food Security in Sub-Saharan Africa. *ITDG Publishing,* London, UK.

DFID (2003), Department for International Development Social Protection, Policy Planning and Implementation, Chapter 20, London, UK.

Dominski, A., Clark, J. and Fox, J. (1992), Building the Sustainable City, Community Environmental Council, Santa Barbara.

Downing, T. E. (1991), Vulnerability to Hunger in Africa- A Climate Change Perspective, Global Environmental Change, 365-380.

Dreborg, K. H., et al., (1996), The Essence of Backcasting, Futures, 28: 813 – 828.

Dsouza, R. M., Becker, N.G., Hall, G. and Moodie, K. B. A. (2004), Does Ambient Temperature Affect Food Borne Disease? *Epidemiology,* 15:86–92.

Economist (2012), Data Everywhere, 25 February, Retrieved 9 December.

Elbehri, A, Anna, S. and Pascal, L. (2013), Biofuels and the Sustainability Challenge: A global Assessment of Sustainability Issues, Trends and Policies for Biofuels and Related Feed stocks. Trade and Markets Division, Food and Agriculture Organization of the United Nations, Rome, Italy.

Emrouznejad, A., Cabanda, E. and Gholami, R. (2010), An alternative measure of the ICT-Opportunity Index, Information and Management, 47: 246–254.

EPA (2009), U.S. Environmental Protection Agency, Washington DC, USA.

Ericksen, P. J. (2008). Conceptualizing food systems for global environmental change research. *Global Environmental Change, 18*(1), 234-245.

European Commission (2001), Summary Innovation Index, DG Enterprise, European Commission, Brussels, Belgium.

Eyben, R. and Ladbury, S. (1994), Popular Participation in Aid Assisted Projects, Why more in theory than practice? in Wright, S and Nelson, S. (eds) Power and Participatory Development: Theory and Practice, Intermediate Technology Publications, London, UK, 192-200.

Fagerberg, J. (2001), In: Lundvall, B., Archibugi, D. (Eds.), Europe at the Crossroads: The Challenge from Innovation-based Growth in the Globalizing Learning Economy. *Oxford Press,* UK.

FAO (2007), Climate Change, Water and Food Security, Technical background document from the Expert Consultation held on 26 to 28 February 2007, FAO, Rome, Italy.

FAO (2008), An Introduction to the Basic Concept of Food Security, Food Security Information for Action Practical Guides, EC-FAO Food Security Program, Rome, Italy.

FAO (2012), Globally almost 870 million chronically undernourished - new hunger report, Food and Agriculture Organization of the United Nations, Rome, Italy.

FAO (2013a), Biofuels and the Sustainability Challenge: A Global Assessment of Sustainability Issues, Trends and Policies for Biofuels and Related Feedstock's, Food and Agriculture Organization of the United Nations, Rome, Italy.

FAO (2013b), Biofuel and Food Security, The High Level Panel of Experts of Food Security and Nutrition, HLPE Report 5, World Food Programme, Rome, Italy.

FAO (2013c), Undernourishment around the World in 2013, The State of Food Insecurity in the World 2013, World Food Programme, Rome, Italy.

FAO (2013d), Information and Communication Technology for Sustainable Agriculture: Indicators from Asia and Pacific, FAO, Bangkok, Thailand.

FAO and WHO (2005), Joint FAO and WHO Food Standards Programme Codex Alimentarius Commission.

Fischer, G, Frohberg, K., Parry, M. L. and Rosenzweig, C. (1996), The Potential Effects of Climate Change on World Food Production and Security, Global Climate Change and Agricultural Production, Food and Agricultural Organization, Rome, Italy.

Fischer, G. (2009), How Can Climate Change and the Development of Bio-energy Alter the Long-term Outlook for Food and Agriculture? Looking Ahead in World Food and Agriculture, Food and Agricultural Organization, Rome, Italy, 96-155.

Fleur, M., Charron, D., Holt, J., Allen, O. and Maarouf, A. (2006), A Time Series Analysis of the Relationship of Ambient Temperature and Common Bacterial Enteric Infections in Two Canadian Provinces, *Int. J. Biometeorol,* 50: 385–391.

Food Security Network (2014), Food Security Network-Healthy Food for All, http://www.foodsecuritynews.com/What-is-food-security.htm .

Fresco, L. O. and Kroonenberg, S. B. (1992), Time and Spatial Scales in Ecological Sustainability, Land Use Policy, 9:155-168.

Georgina, M. and Johnathan, E. M. (2007), The 2010 Biodiversity Indicators: Challenges for Science and Policy, Conservation Biology, 21(6):1406–1413.

GFAR (2009), A Mind Map for ICT in Agriculture, A Global Forum for Agricultural Research.

Global Community (2004), Measurement of Sustainable Development, www://globalcommunitywebnet.com/globalcommunity/measurementofsd.htm, last accessed: 2014.07.08.

Godfrey H.C.J, Beddington JR, Crute I.R et al (2010), Food Security: The Challenge of Feeding 9 Billion People, Science, 327:812-818.

Gornall, J. et al., (2010), Implications of climate change for agricultural productivity in the early twenty-first century, Philosophical Transactions of the Royal Society, Biological Sciences 365(1554): 2973-2989.

Gowda, M. J. C. and Jayaramaiah, K. M. (1998), Comparative Evaluation of Rice Production Systems for their Sustainability, AgricEcosyst Environ, 69:1–9.

GRI (2002), Global Reporting Initiative, Sustainability Reporting Guidelines 2002 on Economic, Environmental and Social Performance, Global Reporting Initiative, Boston, USA.

Guenno, G. and Tizzi, S. (1998), The Index of Sustainable Economic Welfare (ISEW) for Italy. Nota Di Lavoro, 5(98).

Hafeez, K., Zhang, Y. and Malak, N. (2002), Determining Key Capabilities of a Firm Using Analytic Hierarchy Process, *Int. J. Prod. Econ.,* 76:39–51.

Hamilton, A., Mitchell, G. and Yli-Karjanmaa, S. (2002), The BEQUEST Toolkit: A Decision Support System for Urban Sustainability, Building Research and Information, 30(2):109-115.

Hanafizadeh, M. R., Saghaei, A. and Hanafizadeh, P. (2009), An Index for Cross-country Analysis of ICT Infrastructure and Access, Telecommunications Policy, 33: 385–405.

Hayami, Y. and Ruttan, V. W. (1971), Agricultural Development: An International Perspective, 367 l

Hayati, D, Zahra, R. and Ezatollah, K. (2010), Measuring Agricultural Sustainability, *Biodiversity, Biofuels, Agroforestry and Conservation Agriculture, Sustainable Agriculture Reviews*, V(5), Springer Dordrecht Heidelberg London, UK.

Hirschheim, R. A. (1989), User Participation in Practice: Experience with Participative Systems Design', in Knight, K. (ed), Participation in Systems Development, Kogan Page, London, UK.

Höjer, M. (1998), Transport Telematics in Urban Systems: A Backcasting Delphi Study, Transpn Rec–D 3(6): 445–463.

Hojer, M. and Mattson, L. G. (2000), Determinism and Backcasting in Future Studies, Futures, 32: 613–634.

Holt C., Fawcett S.B., Schultz JA, Berkowitz, B. Francisco V. And Wolf T. (2013), The Community Tool Box, Work Group for Community Health and Development at the University of Kansas, *http://ctb.ku.edu/en, last accessed: 2014/08/23*

Holzmann, R. and S. Jorgensen. (2000),"Social Risk Management: A New Conceptual Framework for Social Protection and Beyond. Social Protection."*Discussion Paper No. 0006.* Washington DC.: The World Bank.

Horio, M. (2007), Renewable Energy and Biomass, Chemistry Today, 2 Sept.: 6-30.

IAASTD (2008), Assessment of Agricultural Knowledge, Science and Technology for Development, Summary for Decision Makers of the Global Report, IAASTD Intergovernmental Lenary in Johannesburg, South Africa.

IChemE (2002), Sustainable Development Progress Metrics Recommended for Use in the Process Industries, The Institution of Chemical Engineers, Davis Building, 165-189 Railway Terrace, Rugby CV21 3HQ, UK.

IFPRI (2012), The Food Security System, A New Conceptual Framework, Oliver Ecker, Clemens Breisinger, International Food and Policy Research Institute, Supported by CGIAR.

Ikerd, J. (1993), Two Related but Distinctly Different Concepts: Organic Farming and Sustainable Agriculture, Small Farm Today, 10 (1): 30–31.

Inaba, A. (2005), Practical Use for LCA, Japan Environmental Management Association for Industry, Tokyo, Japan, 314.

Ingram, J., Ericksen, P. and Liverman, D. (2010), Food Security and Global Environmental Change, Earthscan LLC, Washington, DC, USA.

IPCC (1990a), Climate Change: The Climate Change Scientific Assessment eds, *Cambridge University Press,* UK.

IPCC (1992), Climate Change, The Supplementary Report, Scientific Assessment eds, *Cambridge University Press*, Cambridge, UK.

IPCC (2001), Climate Change 2001: Impacts, Adaptation & Vulnerability, Contribution of Working Group II to the Third Assessment Report of the Intergovernmental Panel on Climate Change (IPCC), *Cambridge University Press,* UK.

IPCC (2007), Climate Change 2007: Impacts, adaptation and vulnerability, Contribution of Working, *Cambridge University Press,* Cambridge, UK.

ITU (2012), International Telecommunication Union, *Measuring the Information Society,* CH-1211 Geneva Switzerland.

JEMAI (1999), Introduction of Practical Use for LCA, Japan Environmental Management Association for Industry, Tokyo, Japan, 194.

Jin, S., Zhang, L. and Feng, G. (2011), Earth's Surface Fluid Variations and Deformations from GPS and GRACE in Global Warming, Geoinformatics, 19th International Conference, *IEEE,* 1-4.

Johannes, W. A., Langeveld, J. D., Herman, V. K. and Foluke, P. M. (2014), Analyzing the Effect of Biofuel Expansion on Land Use in Major Producing Countries: Evidence of Increased Multiple Cropping, Biofuels, Bioproducts and Biorefining, 8(1): 49–58.

Jones, P. G. and Thornton, P. K. (2003), The Potential Impacts of Climate Change on Maize Production in Africa and Latin America in 2055, *Global Environ. Change,* 13: 51–59.

Kajikawa, Y. (2008), Research Core and Framework of Sustainability Science, *Sustain. Sci.*3: 215–239.

Kidd, C.V. (1992), 'The evolution of sustainability', *Journal of Agricultural and Environmental Ethics*, 5(1):1–26.

Kleemann, L. (2013), An Overview and Discussion of Solution Proposals for Sustainable Agriculture and Food Security in Sub-Sahara Africa, Sustainable Agriculture Research, 2(4).

Kneen, B. (1989), From Land to Mouth: Understanding the Food System, NC Press, London, UK

Konaka, T. (1997), Bio-production Systems Engineering, shin-Norinsha, Tokyo, Japan, 225 (Japanese).

Konuma, H. (2014), Status of World Food Security and Its Future Outlook, and Role of Agricultural Research and Education, Ag-ESD Symposium 2014, Premium Agriculture and Food Project, University of Tsukuba, Japan

Kovats, R. S., Edwards, S. J., Hajat, S., Armstrong, B. G., Ebi, K. L. and Menne, B. (2004), The Effect of Temperature on Food Poisoning: A Time-series Analysis of Salmonellosis in Ten European Countries, *Epidemiol Infect*,132: 443–453.

Krajnc, D. and Glavic, P. (2005), A Model for Integrated Assessment of Sustainable Development, *Resour. Conserv. Recycl.,* 43:189–208.

Lapola, D. M., et al., (2010), Indirect Land-use Changes Can Overcome Carbon Savings from Biofuels in Brazil, *Proceedings of the National Academy of Sciences,* USA, 107(8).

Lee, J., Cho, H., Choi, B., Sung, J., Lee, S. and Shin, M. (2000), Life Cycle Assessment of Tractors, *Int. J. LCA*, 5(4): 205-208.

Limits to Growth (1972), Donella H. Meadows, Dennis L. Meadows, Jorgen Randers and William @. Behrens III, (1972) Limits to Growth, New York, New American Library.

Lovins, A. B. (1976), Energy Strategy: the road not taken? *Foreign Affairs,* 55: 63–96.

Lundin, U., (2003). Indicators for Measuring the Sustainability of Urban Water Systems—a Life Cycle Approach, PhD Thesis, Department of Environmental Systems Analysis, Chalmers University of Technology, Göteborg, Sweden.

Macgillivray, A. (1996), How a Salmon Can Save the World, Green. *Futures,* 2:44-45.

Maheshwari, P. and Kovalchuk, I. (2014), Genetic Engineering of Oilseed Crops, Biocatalysis and Agricultural Biotechnology, Trait Introduction Methods and Innovation Platforms in Plant Biotechnology, 3(1):31–37.

Mayer, I. (1997), Debating Technologies: A Methodological Contribution to the Design and Evaluation of Participatory Policy Analysis, Ph.D. Thesis Tilburg University, In Quist, J., and Vergragt, P. 2006. Past and Future of Backcasting: The Shift to Stakeholder Participation and a Proposal for a Methodological Framework, Futures, 38(9): 1027-1045.

McLaughlin, A., Mineau, P. (1995),The Impact of Agricultural Practices on Biodiversity, Agriculture, Ecosystems and Environment, 55(3): 201–212.

McNamara, K., Belden, C., Kelly, T., Pehu, E., and Donovan, K. (2011). Overview of ICT in Agriculture: Opportunities, Access, and Cross-Cutting Themes. In *ICT in Agriculture*. Washington, D. C.: The World Bank, InfoDev and ARD, pp 1–14.

Menrolla, J. A. (1989), A Methodology for Ranking Countries According Relative Food Insecurity Presented at the A.I.D., Economists Conference November 12-17, Washington, D.C, USA.

Miao, T., Qingli, X. and Zhongnan, H. Y. (2013), Discussion on the Application of Cloud Computing in Agricultural Information Management, University of Economics and Law, Wuhan, China.

Michael, G. (2000), Community Informatics: Enabling Communities with Information and Communications Technologies, IGI Publishing Hershey, PA, USA.

Millennium Ecosystem Assessment (2005), *Ecosystems and Human Well-being: Biodiversity Synthesis*. World Resources Institute, Washington, DC, USA.

Mochizuki, K. and Sakoda, A. (2005), Construction of Biomass Town Simulator, Production Research 56(2): 148-151 (Japanese).

Morrison, D. F. (1967), Multivariate Statistical Methods, New York, *McGraw-Hill,* USA.

Morse, S. (2004), Indices and Indicators in Development, Earthscan, UK.

Mulder, H. A. J. and Biesiot, W. (1998), Transition to a Sustainable Society: A Backcasting Approach to Modeling Energy and Ecology (e-book), Edward Elgar Publishing, Inc. Massachusetts, USA.

Müller, L. M. (2009), Organic Agriculture and Carbon Sequestration: Possibilities and Constraints for the Consideration of Organic Agriculture within Carbon Accounting Systems, Natural Resources Management and Environment Department, FAO, Rome, Italy.

Munasinghe, M. and Douglas, G. (2007), Multi-criteria Analysis in Environmental Decision-making, Encyclopedia of Earth, Environmental Information Coalition, National Council for Science and the Environment, Washington, DC, USA.

Nadeau, A. and Rudgard, S. (2007). Information Management Resource Kit (IMARK) - Improving information management in developing countries through e-learning. In T. Bastiaens & S. Carliner (Eds.), Proceedings of World Conference on E-Learning in Corporate, Government, Healthcare, and Higher Education 2007: 1149-1154.

Nakata, T., Kubo, K. and Lamont, A. (2005), Design for Renewable Energy Systems with Application to Rural Areas in Japan, *Energy Policy*, 33: 209-219.

Narodoslawsky, M. and Krotscheck, C. (2004), What Can We Learn from Ecological Valuation of Processes with the Sustainable Process Index (SPI): the case study of energy production systems, *Journal of Cleaner Production,* 12: 111-115.

NEDO (2003), Guidebook for Micro hydropower, NEDO, Japan, 138 pp.

NEDO (2008), Guide Line for Regional New Energy Vision and Energy Conservation, NEDO, Japan, http://www.nedo.go.jp/library/pamphlets/ZZ_pamphlets_08_3dounyu_vision_sakutei_index.html, browsed on Sep. 26, 2010.

Nicoletti, G., Scarpetta, S. and Boylaud, O. (2000), Summary Indicators of Product Market Regulation with an Extension to Employment Protection Legislation, Economics Department Working Papers no. 226, ECO/WKP (99)18.

Noguchi, R. and Misumi, M. (2007), Proposal of Rural Energy Simulator Considering Competitive Production between Food and Renewable Energy, Agricultural Information Research 16(3): 158-170 (Japanese).

Noguchi, R. and Takahiro, S. (2008), Consideration of Energy Consumption and Energy Efficiency in Mechanized Rice Production System by using Inventory Analysis, Agricultural Information Research, 17(1): 20-30.

Nugroho, A. S. and Noguchi, R. (2014), Annual Conference of Society of Agricultural Information, Tokyo University, Tokyo, Japan, May 14 – 15.

OECD (1993). OECD core set of indicators for environmental performance reviews. OECD Environment Monographs No. 83. OECD. Paris.

OECD (1997), Environmental Indicators for Agriculture, Concepts and Framework, OECD Publication, Paris, France, 2.

OECD-FAO (2009), Food and Agriculture Organization of the United Nations, Agricultural Outlook 2009-2018, Rome, Italy.

Oliver, M. B. and Brooks, D. B. (2005), The Soft Part for Water in a Nutshell, Friends of the Earth Canada, Ottawa, the POLIS Project on Ecological Governance, University of Victoria, Victoria, Canada.

Pearce, D. and Turner, R. K. (1990), Economics of Natural Resources and the Environment, Harvester Wheat Sheaf Hemel Hempstead, Ch.10.

Petraq, J. P., Panos, M. P. and Antonio, M. (2009), Data Mining In Agriculture, University Of Florida, Gainesville, Fl, USA.

Pimentel, D. and Pimentel, M. (1996), Food, Energy and Society, Univ. Pr of Colorado, USA, 363.

Pingali, P., Alinovi, L. and Sutton, J. (2005), Food Security in Complex Emergencies: Enhancing Food System Resilience, Disasters, 29(1): S5-S24.

Pitcher,T. J and Hart, P. J. B. (1982), Fisheries Biology, *Chapman and Hall,* London, UK.

Population Division, United Nations (2001), United Nations, Department of Economic and Social Affairs, Monitoring Global Population Trends

Porter, M. and Stern, S. (1999), The New Challenge to America's Prosperity: Findings from the Innovation Index, Council on Competitiveness, Washington, DC, USA.

Quist, J. and Vergragt, P. (2006), Past and Future of Backcasting: The Shift to Stakeholder Participation and a Proposal for a Methodological Framework, Futures, 38(9): 1027-1045.

RAMP (2014), Regional Aquatics Monitoring Program, http://www.ramp, last Accessed on November 1, 2014.

Reijntjes, C., Bertus, H. and Bayer, A. (1992), Farming for the Future: an introduction to low external input and sustainable agriculture, *Macmillan,* London, UK.

Riely, F., et al., (1999), Food Security Indicators and Framework for Use in the Monitoring and Evaluation of Food Aid Programs. Food and Nutrition Technical Assistance Project (FANTA): Washington, DC, USA.

Robinson, J. (1982), Energy Backcasting: A Proposed Method of Policy Analysis, Energy Policy, 10: 337–344.

Robinson, J. (2003). Future subjunctive: backcasting as social learning. *Futures*, *35*(8), 839-856.

Rogers, E. (1995), Diffusion of Innovations (4th ed), Free Press, New York, USA.

Royal Society of London (2009), Reaping the Benefits: Science and the Sustainable Intensification of Global Agriculture.

Ruiz-Garcia, L. and Lunadei, L. (2011), The Role of RFID in Agriculture: Applications, Limitations and Challenges, Computers and Electronics in Agriculture, 79(1): 42-50.

Rutgers (2014), From Farm to Fork, http://njaes.rutgers.edu/health/farmtofork. asp, Last Accessed: 2014.11.7

Ryan, B. (2014), Harvesting the Power of Big Data in Agriculture, 2014. http://www.engr.wisc.edu/june11-big-data-agriculture,
Last Accessed on 2014.09.16.

Saaty, T. L. (1995), Decision Making for Leaders: the analytic hierarchy process in a complex world. RWS Publications, Pittsburgh, Pennsylvania, USA.

Sala, S. (2007), Information and Communication Technologies for Climate Change Adaptation with a Focus on the Agricultural Sector, Università degli Studi di Milano, Crop Production Department, *FAO WorkingPaper,*Rome, Italy.

Scaramozzino, P. (2006), Measuring Vulnerability to Food Insecurity, *ESA Working Paper,* 06-12.

Sen, A. (1988), The Concept of Development, Handbook of Development Economics Amsterdam: North Holland, 1.

Senge, P., Ross, R., Roberts, C., Smith, B. and Kleiner, A. (1994), The Fifth Discipline Field Book: Strategies and Tools for Building a Learning Organization, Nicholas Brealey, London, UK.

Singh, R. K., Murty, H. R., Gupta, S. K. and Dikshit, A. K. (2009), An Overview of Sustainability Assessment Methodologies, Ecological Indicators, 9(2): 281-299.

Sonka, S. (2014), Big Data and the Ag Sector: More than Lots of Numbers, International Food and Agribusiness Management Review, 17: 1-19.

Steen, P. and Akerman, J. (1994), Syntesav studier over omstallning avenergi-ochtransportsystemeniSverige, A report from the Swedish Governmental Committee on Climate, 28(9): 813-828

Steffen, W., Sanderson, A., Tyson, P.D., Jager, J., Matson, P.A., Moore, III, B., Oldfield, F., Richardson, K., Schellnhuber, H.J., Turner, II, B.L., Wasson, R.J. (Eds.), 2003. Global Change and the Earth System: A Planet Under Pressure. SpringerVerlag, Berlin, New York.

Stewart, T. J., Janssen, R. and Herwijnen, M. V. (2004), A Genetic Algorithm Approach to Multi Objective Land Planning, Computers and Operations Research, 31: 2293-2313.

Stockley, D. (2003), E-learning Definition and Explanation, Retrieved September 17, 2007.

Streeten, P. (1994), Human Development: Means and Ends, The American Economic Review, 84(2):232-237.

Suhong, L., Visich, J. K., Basheer, M. K. and Zhang, C. (2006), Radio Frequency Identification Technology: Applications, Technical Challenges and Strategies, Sensor Review, 23: 193–202.

Ten Brink, B. J. E., Hosper, S. H. and Colijn, F. (1991), A Quantitative Method for Description and Assessment of Ecosystem, The AMOEBA Approach, *Marine Pollution Bulletin,* 23:265-270.

Ting, K. C. (2009), Engineering Solutions for Biomass Feedstock Production, Resource, April/May 2009.

Ting, K. C., Abdelzaher, T., Alleyne, A. and Rodriguez, L. (2011), Information Technology and Agriculture: Global Challenges and Opportunities, The Bridge, National Academy of Engineering, Fall 2011 Agriculture and Information Technology.

Tochigi Branch Office (2006). Annual statistics of agriculture, forestry and fisheries in Tochigi prefecture, Kanto regional agricultural administration office, (Japanese).

Todorov, V. and Marinova, D. (2010), Modeling Sustainability: Mathematics and Computers in Simulation, 81:1397–1408.

UN (2008). World Population Prospects, the 2008 Revision, United Nations, New York, USA.

UN (2009), United Nations, World Water Development Report

UNDP (2001), Human Development Report, *Oxford University Press,* New York, USA.

UNDP Human Development Report (2007), Fighting Climate Change, Human Solidarity in a Divided World, United Nations Development Programme, New York, USA.

UNEP (2011), *Biofuels Vital Graphics, Powering a Green Economy*, UNEP, Division of Technology, Industry and Economics, United Nations Environment Programme.

United Nations World Food Program (2009), The World Food Programme's 2009 Annual Report, UN, USA.

Van, K. M. (2004), Debating Climate Change: A Study of Stakeholder Participation in an Integrated Assessment of Long-term Climate Policy in the Netherlands, PhD thesis, Free University, Amsterdam. Netherland.

Wakeyama, T. and Ehara, Y. (2009), Assessment of Renewable Energy by Using GIS : A Case Study of Unzen City, *Journal of the Japan Institute of Energy,* 88: 58-69 (Japanese).

WCED (1987), Our Common Future, World Commission on Environment and Development, Oxford University Press, Oxford, UK.

Webster (1995), Mirriam Webster's Collegiate Dictionary, Mirriam Webster, Chicago, USA.

Webster, P. (1999), The Challenge of Sustainability at the Farm Level: Presidential Address, *J Agric Econ* 50(3): 371–387.

Weersink, A. (2001), Multifunctionality: Applying the OECD Framework, a Review of Literature in Canada. Directorate for Food, Agriculture and Fisheries.

WHO (2007), World Health Organization, International Radon Project Survey on Radon Guidelines, Programmes and Activities, WHO, Geneva

WHO (2014), Trade, Foreign Policy, Diplomacy and Health, http://www.who.int/trade

Wieland, R., Voss, M., Holtmann, X., Mirschel, W. and Ajibefun, I. (2006), Spatial Analysis and Modeling Tool (SAMT), 1.Structure and possibilities, *Ecological Informatics,* 1: 67–76.

World Bank (2010), World Development Report 2010: Development and Climate Change. Washington, DC, USA.

World Economic Forum (2014), Global Information Technology Report, Rewards and Risks of Big Data, Geneva, Sitzerland.

World Water Assessment Programme (2009), The United Nations World Water Development Report 3: Water in a Changing World, Paris, UNESCO and London: Earthscan, 112.

Yamamoto, H., Yamaji, K. and Fujino, J. (2000), Scenario Analysis of Bioenergy Resources and CO_2Emissions with a Global Land Use and Energy Model, Applied Energy, 66:325-337.

Yunlong, C. and Smit, B. (1994) Sustainability in Agriculture: a general review, Agriculture, Ecosystems and Environment, 49: 299-307.

Zabel, R.W., Harvey C.J. Katz S.L. Good T.P. Levin P.S (2003), Ecologically Sustainable Yield, *American Scientist,* 91(2): 150-157.

Zadeh, L. A. (1965), Fuzzy Sets, Information Control, 8: 338-353.

Zgurovsky, M. Z. (2007), Sustainable Development of Global Simulation: Opportunities and Treats to the Planet, *Ruus J. Earth Sciences,* 9:1-16.

Zhu, X., Clements, R., Haggar, J., Quezada, A., & Torres, J. (2011). *Technologies for Climate Change Adaptation-Agriculture Sector.* Danmarks Tekniske Universitet, Risø Nationallaboratoriet for Bæredygtig Energi.

ABOUT THE AUTHORS

Tofael Ahamed is an Associate Professor and Faculty member of Life and Environmental Sciences at the University of Tsukuba, Japan. Dr. Ahamed received his PhD in Bioproduction and Machinery from the University of Tsukuba. His teaching interests include agricultural research methodology, the concepts of a sustainability index, sustainable agriculture and food security at the graduate program of International Agricultural Sciences at the University of Tsukuba. From the engineering solutions perspective, Dr. Ahamed teaches instrumentation, bioproduction systems, precision agriculture technology, farm machine design and advanced mathematics for life and environmental sciences.

Ryozo Noguchi is an Associate Professor and Faculty member of Life and Environmental Sciences at the University of Tsukuba, Japan. He received his Ph.D. in Agricultural Machinery from the Department of Agricultural Engineering at Kyushu University. Dr. Noguchi teaches Energy Systems Engineering, Agricultural Production Systems, and Bio-resource Utilization. His research focuses on biomass and renewable energy utilization in rural areas and agricultural informatics with decision-support systems established through simulation and modeling.

Tomohiro Takigawa is a Professor and Faculty member of Life and Environmental Sciences at the University of Tsukuba, Japan. He received his PhD in Automation and Post-Harvest Processing from the University of Tsukuba. Dr. Takigawa teaches classes in advanced agricultural machinery, agricultural robotics and research ethics. His research focuses on developing autonomous agricultural machinery, tractor-trailer dynamic modeling for off-road equipment, and nonlinear control systems for autonomous vehicles.

Index